Afterlives of Data

Afterlives of Data

LIFE AND DEBT UNDER CAPITALIST SURVEILLANCE

Mary F. E. Ebeling

UNIVERSITY OF CALIFORNIA PRESS

University of California Press
Oakland, California

Library of Congress Cataloging-in-Publication Data

Names: Ebeling, Mary F. E., author.
Title: Afterlives of data : life and debt under capitalist surveillance / Mary F. E. Ebeling.
Description: Oakland, California : University of California Press, 2022. | Includes bibliographical references and index. | Identifiers: LCCN 2021058916 (print) | LCCN 2021058917 (ebook) | ISBN 9780520307728 (cloth) | ISBN 9780520307735 (paperback) | ISBN 9780520973824 (epub)
Subjects: LCSH: Medical records—Access control—Economic aspects—United States. | Debt—Political aspects. | BISAC: SOCIAL SCIENCE / Privacy & Surveillance (see also POLITICAL SCIENCE / Privacy & Surveillance) | TECHNOLOGY & ENGINEERING / Social Aspects
Classification: LCC RA976 .E28 2022 (print) | LCC RA976 (ebook) | DDC 610.285—dc23/eng/20220121
LC record available at https://lccn.loc.gov/2021058916
LC ebook record available at https://lccn.loc.gov/2021058917

Manufactured in the United States of America

30 29 28 27 26 25 24 23 22
10 9 8 7 6 5 4 3 2 1

For all of you caretakers and dreamers who continue to draw upon a radical love and rage to fight for those whom we have lost. May we never go back to normal.

Contents

Acknowledgments

The essayist and humorist Fran Lebowitz, well known for her decades-long writer's block, sums up her abiding aversion to writing thusly: "Contrary to what many of you may imagine, a career in letters is not without its drawbacks—chief among them the unpleasant fact that one is frequently called upon to sit down and write" (Lebowitz 2011, 152). Never have truer words spoken to my own solitary agony.

Writing any book is such a long and lonely process. And writing an academic book, one that only a few hearty souls will be able to bear to read (but thank you for reading this one!), almost seems like a pointless exercise in self-loathing and self-imposed punishment. To focus on writing a book during a traumatizing year of desperation, anxiety, and loss spurred on by the pandemic, by the uprisings against fascism and social injustice, and by the insurgencies against democracy, was exceptionally challenging. Not only because it was hard to concentrate and think coherently during such an unprecedented time, but also because it felt to me like such an indulgent and selfish act. Yet in spite of my feelings about the drudgery of writing and my guilt about not doing more to help people vote or eat or stay safe, the book has, miraculously, been written. But books are never written alone, even if writing them can be a lonesome exercise. I have the great privilege to be in community with other scholars, colleagues, friends, and family who are much smarter and wiser than I am, and who all, in one way or another, contributed to this book's existence. In all honesty, this book would not have been possible without the guidance, insight, encouragement, generosity, and patience of so many others. I am deeply grateful that all of you are in my life. But bear with me as I acknowledge and thank those specific people, and some institutions, that helped me along the way to complete this book.

Some of the core ideas in *Afterlives of Data* were developed while I was a resident fellow at the Wolf Humanities Center at the University of Pennsylvania during the 2017–18 academic year. That year's theme was "afterlives," and our weekly seminars had a direct influence on the core premise of the book: that our personal health and financial data live multiple and disparate afterlives, beyond their initial usefulness or original purpose. The theme was led by the phenomenal feminist translator and classical studies scholar Emily Wilson, with support from literature scholar and center director James English. Both deserve my thanks for including me in that year's cohort. I was inspired, invigorated, and profoundly moved by the brilliance and dedication of each and every participant. I continue to consider myself lucky to have learned about the European medieval concept of the soul, anchoress Julian of Norwich's *Revelations of Divine Love*, the rare books stolen from Jewish families by the Nazis and rescued by archivists, and scientific investigations into reincarnated lives. Thanks to all of the fellows who inspired me every Tuesday, including James Bergman, Jacob Doherty, Adam Foley, Darin Hayton, Christopher Lee, Colin Williamson, Kathy Peiss, Simon Richter, Salamishah Tillet, Fatemah Shams, Delia Wendel, Huda Fakhreddine, Sierra Lomuto, Orchid Tierney, and Sara Ray. I was particularly moved by the work of Delia and Fatemah, both of whom helped to excavate and memorialize collective traumas, and to bring dignity to those who have survived unspeakable loss. Thanks to Meg Leja especially for our conversations about the intersectional and gendered sacrifices that are expected of us in academia, and for sharing the Eagles' Super Bowl win with me. I will never be a football fan but I will always love the Eagles. Particular gratitude is owed to Projit Mukharji for his interest in my work, and to S. Pearl Brilmyer for the close reading that she brought to my writing. I must acknowledge and thank Sarah Milinski, not only for her organizational support, but perhaps more significantly for our shared love for Prince.

Another seminar that influenced some of this book, in particular chapter 6, was the Seminario sobre la deuda, held at Beta Local in San Juan, Puerto Rico, during June 2016, and organized by the brilliant powerhouse of a filmmaker, an artist, and a codirector, Sofía Gallisá Muriente. There I learned from activists, artists, filmmakers, economists, journalists, poets, performers, musicians, and politicians about how the Puerto Rican national debt is a crisis of colonization and US imperialism that shapes every aspect of life on the archipelago. My return to Beta Local in 2019 to interview some of those that I had met in the debt seminar, was only possible because of the generosity and support of Beta's codirectors, Sofía, Pablo Guardiola, and Michael Linares. Thank you all for the space to think, for the engaging conversations in the Callejón bar, and for all the enthusiasm for our collaborations, especially when it comes to making Ethiopian cuisine. Thanks also goes to Alix Camacho, Gisela Rosario Ramos Rosa Vazquez, Bea Santiago Muñoz, Marina Reyes Franco, Monxo Lopez, Dave Buchen, Tony Cruz, Macha Colón, Marichi Sharron, Oli, and Dziga. And I must acknowledge Blanca Rosa Rovira Burset and her lovely elderly students at the El Escambrón water aerobics class. Thanks for inviting me to join in and for singing the songs of your shared childhoods. You all gave me a cherished memory of San Juan. I can't wait to join you again.

Many thanks are due to some of the phenomenal data scientists whom I have met along the way and who patiently helped to translate their work for me. Your kindness, wisdom, and generosity are deeply appreciated Aixa Guo, Anuj Shahani, Sige Diagne, Randi Foraker, and Po-Yin Yen.

A special mention of thanks is owed to several people scattered across Europe for inviting me to join in seminars and conversations. These include everyone in the Data Studies and Knowledge Processing Group within University of Exeter's EGENIS Centre, led by Sabina Leonelli, with an extra helping of gratitude to Niccolò Tempini, Gregor Halfmann, and Sabina herself. Thanks to Susan Kelly, also at the University of Exeter. I am grateful to Federica Lucivero, Bronwyn Perry, and Barbara Prainsack for bringing me to King's College London and to the Ethox Centre at the University of Oxford; these visits meant a lot to me, especially at a time of great uncertainty and anxiety. Gratitude beyond words goes to Christine Aicardi: thank you for trusting me with the memories that you keep close to your heart. You really are one of the bravest women I have ever known. Many thanks also to the graduate students and participants at the Centre de sociologie de l'innovation (CSI) at Mines ParisTech who honored me with a seminar; some of your insights and questions made it into this book. I want to thank especially Evan Fisher, Loïc Roim, Kevin Mellet, Jérôme Denis, Thomas Beauvisage, and Yasemin Ozden Charles. And to Rob Meadows and Christine Hine, I deeply appreciate that you always include me on the research team at the University of Surrey. After the pandemic is over, we will have that meeting. Sue Venn, I have not forgotten what you shared with me years ago, and I hold that knowledge with respect. Albaqir alafif Mukhtar, your bravery and strength know no bounds, as does your kindness and warmth. Deb Carson, thanks for everything, really. Your enduring friendship means so much to me. I can't wait to return to the Medway towns for Sweeps and to see what's growing in the allotment.

This book absolutely would not have been possible without the support that I received from Drexel University's Faculty Scholarly and Creative Activity Award during 2019–20. This grant allowed me to conduct research in the field sites that are discussed in the book, and more importantly, it enabled me to hire the phenomenal Dina Abdel Rahman as my research assistant. Dina's abundant creativity, passion, and brilliance shines brightly throughout the book. Dina tirelessly tracked down literature, analyzed content, bounced ideas around with me, and contributed fresh perspectives and insights, all of which was informed by her own experience of working with health data in emergency medicine research. Dina, none of this would have been possible without you. Thank you for accompanying me on this long and at times thorny path. I look forward to calling you doctor soon. Much appreciation goes to Rachel Koresky, department administrator at Drexel University, for all of her help with managing the grant.

Sections of *Afterlives of Data* are the result of two previously published articles that were greatly expanded upon and altered here. Parts of chapters 2 and 3 were published in 2019 as "Patient Disempowerment through the Commercial Access to Digital Health Records," and appeared in *Health* 23 (4): 385–400. The second half of chapter 4 and a section of Chapter 5 were published in 2018 as "Uncanny Commodities: Policy and Compliance Implications for the Trade in Debt and Health Data," in *Annals of Health Law and Life Sciences* 27 (2): 125–47. Both articles benefited immensely from the anonymous reviewers who provided feedback and from the editors at each of the journals. A word of thanks is also owed to the editors who supported me through the book-writing process, including Ellen Goldstein, Noreen O'Connor-Abel, and Jennifer Jefferson. I am thrilled that the phenomenal artist Julianna Foster allowed me to use a photograph from her Swell series as a visual representation of the book on the cover; thank you, Jules. And my deepest gratitude goes to Maura Roessner, my editor at the University of California Press, whose unfailing belief in this book, and in my ability to finish it, sustained me through the pandemic years. Madison Wetzell, editorial assistant at the University of California Press, deserves a note of thanks as well. Any errors or mistakes are solely mine.

Finally, my deep appreciation is due to all of you who are my loves, my companions and co-celebrants of joy and participants in gigglefests, and my sustainers through hard and lonely times. All of you who, by my good fortune, are in my life and continue to inspire me. These special souls include the kickass poets Jena Osman, Ren Ellis Neyra, and Tsitsi Jaji; the kickass artists Maura Cuffie, Rea Tajiri, Hye-Jung Park, and Maria Dumlao; and other category-defying people who kick ass, like Zoe George, Chiming Yang, Sirat Attapit, Victor Peterson II, Franklin D. Cason, Tembo, Debjani Bhattacharyya, Vincent Duclos, Souleymane Bachir Diagne, Brooke Dorothy, Adam Knowles, Janet Golden, Asta Zelenkausaite, Siddiq Abdal Hadi, Michele White, Timothy Lancaster Hunt-Cuticchia, Jenny Hoare, Matt Hunt, and cousins Cecile and Tony. Last but not least of those who kick ass, Ripley. A special thanks to the kick-ass *chingonas* at Juntos, my happy place, where you all helped to keep everything in perspective, including Erika Guadalupe Nuñez, Rebecca Gonzalez, Guadalupe Mendez, and Nikola Boskovic. Appreciation also goes to Olivia Ponce for dancing with me in front of the convention center. Thanks to all of you who sacrificed personal safety, and sometimes your lives, to care for one another and unknown others, and to undo the workings of the death cult. Much love and gratitude to my family of caretakers including Louis

Portlock and Elaine Prewitt, and Mike Sherwood. And to the eternal loves of my life and keepers of my heart, Tim Portlock and Mashashe Savannah Ward.

I wrote most of this book while living on the land of the Lenni Lenape, and I gratefully acknowledge the ancestors of this place, past, present and future. I also acknowledge the keepers of Cahokia, and the keepers of rivers and oceans.

I hope that my next book will be purely fiction.

Introduction

DATA LIVES ON

Ana V. Sanchez is dead. Writing these words floods me with many emotions: sadness, anger, hopelessness. Above all, I am filled with shame. Ana and I spoke perhaps three times before she died of brain cancer several years ago. Ana was not my cousin or sister, or even a very close friend. She was more of an acquaintance, a vaguely familiar face that I would sometimes see in a crowd at a party.

"We met at that art opening in 2009, right?" Not much else connected Ana and me. The last time we saw each other, she had to ask me to remind her of my name. It saddened me when she asked because I detected a slight slur in her question. "Mary," I reminded her. And she beamed. "Maria? Oh, yes, yes, Mary."

I first met Ana in 2009, at an art opening, where we instantly hit it off. Ana, who had a piece in the show, was a fiber artist, and much of her work involved large installations—of luxuriant red satin sewn into hanging walls of peony petals, or of cotton balls stitched together with fine red thread to make clouds that seemed to bleed. In her hands, lifeless fabric was transformed into verdant invocations of life. Thinking back to the first time I met Ana, I recall her face radiating an inner warmth and kindness. Her chestnut hair was long and glossy, which seemed to mimic the walls of cascading satin petals. As we circulated in the gallery to look at each artist's work, we talked. I learned that Ana had grown up in Spain, and that while some of her relatives, like her dad, lived in Puerto Rico, she was by herself in Philadelphia, far away from her family. I assumed that Ana found herself, as I did, living in a society that valorizes individualism and self-reliance, and that tends to see personal hardship, especially in regard to health or debt, as a personal failing, rather than as a brutal symptom of systemic dispossession and injustice.

The second time I saw Ana, it was at the opening of a group show. This time Ana was wearing a white shirt with a little black bow tie clipped onto the collar, black trousers, and black shoes. Her lush hair was pinned back into a neat bun. Her art-making alone could not support her financially, and her catering jobs were often for art openings that featured the work of her friends and colleagues in the tight-knit art community in Philadelphia. Like so many American workers, she learned to live with contingent contracts, precarious income, and little stability. Ana learned to survive, barely, with no health insurance. We only saw each other briefly that night, when she offered me an hors d'oeuvre, a miniature quiche, from her tray as an excuse to say hello and talk for a few minutes.

The last time I saw Ana, it was after her third surgery to remove the brain tumor, at yet another art opening. This time, she was just visiting, not serving food or drinks. Her hair was now brittle and completely gray. It was still long, which enabled Ana to pin it over the side of her head where a five-inch scar snaked around her skull. Ana saw me, and in an instant, tears welled up in her eyes. She had a catch in her throat as she thanked me for my donation to her YouCaring site, an online crowd-sourced fundraising campaign for medical debt. I had given fifty dollars about a year earlier to help pay for one of her brain surgeries. As far as I knew, the hospital that was treating her cancer hadn't helped her to set up the online campaign, which is one of several third-party financing services that medical practices now offer insured, underinsured, and uninsured patients to help them pay for their treatment costs. That evening was the last time I saw Ana alive. She died later that year, in November, just a few months shy of her forty-first birthday.

Cancer may be among the costliest medical conditions to treat, and its treatment places a huge financial strain on the individual patient, their families, and the healthcare system broadly (Mariotto et al. 2011; Park and Look 2019). Even for those who have insurance, the complex billing processes and co-pays—costs associated with procedures and drugs that insurance does not cover—and other out-of-pocket expenses force many into bankruptcy. This treatment-related devastation has been named "financial toxicity" among cancer healthcare researchers, and is related to lower-quality care, poorer outcomes, and higher mortality rates (Zafar 2016; Zafar and Abernethy 2013). The financial toxicity associated with oncological medical treatments makes patients sicker and less likely to survive. In consideration of these overwhelming financial burdens for patients and their families, but also to ensure that medical providers are paid, many hospitals and cancer treatment clinics now offer financial counseling. This counseling can include advising patients to look to alternative treatment financing, and helping to promote a patient's crowdfunding campaign to pay for cancer treatment (Berliner and Kenworthy 2017; Cohen et al. 2019). These days, too, when a patient's account goes into arrears with a medical provider, hospitals turn to debt collectors, and to the courts, to go after their patients for these unpaid debts. These patient debtors may even end up in prison.[1]

Since I didn't know Ana that well, I could only speculate as to the financial measures she had already exhausted, from maxing out her credit cards to borrowing from friends and family. I imagined that Ana's medical debt was considerable if she had turned to online crowdfunding to pay for her treatment. Ana's debt—information that would have embodied, and been marked by, the traumas and fears that accompany such financial liabilities—was no doubt fed into credit-risk algorithms as simple points of data, numbers that would have tanked her credit score (Dean et al. 2018; Zeldin and Rukavin 2007).[2] Before that, though, as a gig worker with no insurance, she had most likely opted out of earlier medical care that might have slowed the progression of her illness, or at the very least, improved the quality of her life.

The last time I saw Ana, all of the unbearable alienation, desperation, and shame brought on by our profit-driven healthcare system condensed into a hardness in our throats. We shared a common subjectivity; we were both subjects in a medical system driven by debt and data, where healthcare is not a right but a privilege.[3]

The shame that I felt that last night that I saw Ana continues to haunt me. I had given her fifty dollars: that's it, next to nothing in the grand scale of funding terminal brain cancer treatment with or without health insurance. I did not give her any of my time or care while she was dying. I didn't bother to write my representatives to demand healthcare as a human right nor did I even join a coalition, such as the Poor People's Campaign or groups advocating for Medicare for All. I couldn't manage to attend even one protest, or a hearing for healthcare financing reform. This angry shame that I carry boils down to this: Ana's gratitude for the charity of strangers, even for a measly fifty dollars delivered by the click of a button, was misplaced in that she was grateful for something that should have been hers by right. Ana died despite that tenuous shred of empathy shared with her, because in the brutal system of Hobbesian healthcare, there is little room for tenderness.

The United States has one of the most expensive healthcare systems in the world. Americans spend close to 18 percent of the gross domestic product (GDP, or total economic output) on healthcare (Centers for Medicare and Medicaid Services n.d.). Because it is technologically advanced and uses cutting-edge innovations such as machine learning and artificial intelligence (AI) diagnostics or the latest gene therapies, this high-cost healthcare system is cited as the best in the world. However, the most expensive does not mean the best, or even simply better, especially when it comes to equity. Many scholars who study healthcare policy in the United States, from healthcare economists to public health researchers, have long noted that these expenditures indicate that our healthcare system not only is inequitable, but also drives inequality (Nunn, Parsons, and Shambaugh 2020). The high costs in American healthcare result not from technological innovations or overutilization in general, but rather from the high prices of services and administrative expenses associated with a predominately for-profit system (Papanicolas, Woskie, and Jha 2018).

The cruelty of Ana's last few months struggling to survive brain cancer belies all the promises made about the high quality and effectiveness of the American healthcare system. These promises continue to go unfulfilled for millions of Anas in a society where shame and gratitude are the currencies transacted among us, instead of mutual commitment and obligation to one another. Why is it that in late capitalism, in for-profit healthcare, the means of exchange always place us at the sharp end of a transactional relationship, one that pits our survival against giving up control over our data and our financial futures? Medical capitalism forces many patients, insured and uninsured alike, into debt, pushing life-saving healthcare out of reach for more Americans and amplifying health and economic disparities. The COVID-19 pandemic has further elucidated and deepened these stark systemic inequities.

Ana passed away in November 2016. Just a few years later, the United States weathered another economic crisis and recession, brought on by a pandemic, as well as an uprising against systemic racism and violence. Unequal social conditions persisting from the 2008 global economic recession worsened, and many Americans remained in danger of getting sick or dying of the coronavirus; or being denied access to care, including the vaccine; or losing their savings, jobs, and homes. While the social inequities in health and wealth seem even worse than they were just a few years earlier, when Ana died, some things have not changed much. The skewed bargains that ensure that corporate interests control and profit from our data in healthcare and in finance persist and flourish. The tech giants, like Amazon and Google, as well as the health insurers and credit-reporting bureaus all reported record-breaking profits during the pandemic (Plott, Kachalia, and Sharfstein 2020).[4]

AFTER DEATH

I always saw a connection between what I imagined happened to Ana—and the possible ways that data about her oncological care and medical debt lived on after she died—and my own story of personal grief. At the time that Ana died, I was struggling with data concerning my own health and medical debt. These data were animated and given a rather phantasmagoric afterlife. For more than five years, a phantom haunted me. Like many ghostly apparitions, my phantom didn't have a body, but I sensed its presence in my day-to-day life. It didn't have a voice, but it spoke just the same. Though it was a complete stranger, my specter was uncannily familiar. It was intimately connected to my body, my health, and my life. The phantom grew to become my kin, my descendant: It was my data revenant. It was a ghostly creature whose body was made from the health, financial, and consumer data that lived on my credit cards in the form of medical debt or in the opaque information that comprised my credit score, in the cookies and browser history on my laptop, and in countless other digital crumb trails that I have left over the years.

You could say that the genesis of my data revenant, as well as this book, began on March 14, 2011—the day I learned that I had had my final miscarriage. That morning my doctor showed me that the cluster of glowing pixels on the ultrasound's monitor, which had vibrated with my baby's heartbeat only a week earlier, had gone still and silent. It was in the afternoon of that gray, early spring day that I met my data baby for the first time. Just like a shapeshifter or, as I came to understand it, a revenant, my data baby took on the shape of my fleshy baby that might have been. Instead of blood, skin, and bones, my data baby was composed of the free samples of baby things and other targeted marketing appeals sent directly into my mailbox or my email inbox for five years after my miscarriage. Where my pregnancy resulted in a miscarriage, my data baby gestated for nine months and was born at term, with no complications and completely healthy. Over those years, my data baby grew from being a newborn, to a toddler, to a preschooler, and finally, the last time that I heard from my data baby, it was getting ready to start kindergarten.

In the depths of my grief back then, I could not have imagined or known how the loss of my baby would eventually lead me to dedicate so much time, emotional labor, and psychic energy to studying something that most of us can't see or touch. At times, we can sense its presence humming in the background of our daily lives, and for some of us, its existence can make itself known, often in disruptive or harmful ways.

What we sensed happening "behind the scenes" has come under the glare of public scrutiny and concern. The *New York Times Magazine*'s 2012 investigative piece "How Companies Learn Your Secrets" uncovered how retailers like Target customize price markdowns to individual consumers by collecting data on in-store and online sales transactions, which they then run through predictive analytics. At several antitrust hearings, members of Congress publicly interrogated Big Tech executives—Alphabet (parent company of Google) CEO Sundar Pichai, Facebook executives Mark Zuckerberg and Sheryl Sandberg, and Twitter CEO Jack Dorsey—

on their industry's mishandling of millions of users' personal information and their platforms' role in generating dangerous pandemic and election misinformation and extremism (Ebeling 2016).[5] For many people data means not revolutionary promises of empowerment and liberation, but surveillance, loss of control and freedom, intrusions, division, and harm. The incremental and skewed bargains that we make concerning our data have detrimental consequences for our health, our finances, and our lives (Pasquale 2015; Zuboff 2019). Data about our lives, in all their varied forms, produced by us, lead existences that run parallel to our own "real life." The data, which often define us in ways that we struggle to recognize, can never represent the fullness of our lives. As I show, these data survive even death.

My research into all the places where our health and financial data live on started in 2014. Spurred on by my grief, anger, and the trauma of being haunted by data, I was determined to hunt down the pieces of my data baby in the disparate databases of marketers and reassemble them to make them whole. This research has defined my scholarship ever since. Through my investigations into how my health data ended up in the hands of marketers, I learned that the data baby was conceived with the purloined data taken from interactions in my day-to-day life. Data brokers harvested these data in places not limited to but including the pharmacy where I purchased over-the-counter ovulation test kits and the prescribed drugs used in in-vitro fertilization protocols, the fertility clinic where my credit card transactions paid for the copays and costs of medical treatment that my health insurance did not cover, and my computer and my phone where metadata about my online behaviors was generated as I visited websites about what I could expect in early pregnancy. I learned that my data went on to have multiple afterlives in all sorts of databases, owned and maintained by companies that I had never heard of, much less did business with. Some of my data were used to inform clinical trials or to boost the published success rates for the fertility clinic (because I did get pregnant, after all), and other bits of my data were segmented into marketing categories and commodified by data brokers, such as Experian plc. One of the largest data brokers, marketing services, and credit bureaus in the United States, Experian is as powerful as Alphabet when it comes to accessing and monetizing our data. However, most Americans know Experian only as a company that produces credit reports and credit scores. In recognition that the handling, trading, and monetizing of private health and financial information by powerful companies like Experian is widespread, in 2000, Congress amended parts of the Gramm-Leach-Bliley Act of 1999 (GLBA) that directly address the privacy and security of personally identifiable information. The legislation originally deregulated the financial services, insurance, and banking sectors to allow for mergers among these sectors; the amendments added privacy safeguards in an effort to protect consumers and patients from harmful cross-sector data exposures.[6] These legislative guardrails, however, did not prevent the consolidation of power over personal, private data into the hands of a few corporations in the financial services and credit information reporting sectors. Because these powerful players control vast amounts of our data, they can also determine outcomes in the high-stakes decision-making processes that use our data in healthcare and personal credit and banking services.

I wanted to find out how a company that I did not have any known direct contact with got hold of my private health data, especially data that, at the time, I naively considered to be mine, or at the very least, co-owned by myself and my doctor. I could not understand how a credit bureau could access information about my health, especially clinical data that most of us believe are protected by special privacy regulations; in particular, by the 1996 Health Insurance Portability and Accountability Act (HIPAA). I interviewed data brokers, data marketers, healthcare professionals (doctors, nurses, surgeons), bioinformaticians, pharmaceutical sales reps, and other patients who were haunted by their own data revenants. Through this process I uncovered how our data, especially information that is interpreted as health data, are harvested, commodified, and made to produce value for those that come to own our data. Taking up the role of a hardboiled detective, I wrote an autoethnographic noir about my investigations into data-marketing surveillance, the biodata economy, and how our medical data come to haunt us, in *Healthcare and Big Data: Digital Specters and Phantom Objects* (Ebeling 2016). With *Afterlives of Data,* I continue and expand upon the research started in my first book. Here my aim is to understand how our traumas, losses, and shame haunt our data, and how we are subjects of the double regimes of debt and data. Those subjectivities of trauma and loss, of debt and dispossession, become embodied as the data assets that produce value for the data economy at large. Often, that value does not return to us, the subjects of data and debt.

A NOTE ON METHODS

At a field site that I was researching for this book, one of my interlocutors and collaborators, Anmei, a nursing informatics researcher, aired her frustrations with a statistical model. She couldn't make the data fit the model, a conundrum I would often hear about from data scientists, and if she forced it, the model would no longer work. "Give me a number, a value, and it is either yes or no, and I can get to the truth," she said. "Qualitative research, sociology, what you do, is too difficult; the truth is not just one thing." Anmei's desire to stand outside the phenomenon under study by objectifying it through numbers is an impossibility for many sociologists, especially those who use qualitative methodologies like ethnography, my preferred method, to understand social phenomena.

In his book *After Method: Mess in Social Science Research* (2004), John Law warned that while "the social" appears to be definite, something discoverable that can be validated scientifically, social realities are fragmentary, multiple, complex, and always locally situated. The trick is to remember that the very social science methods developed to "know" these phenomena simultaneously help to create the same social realities that are being studied. Law noted that amid the ongoing debates about the nature of social reality, "the task is to imagine methods [that] no longer seek the definite, the repeatable, the more or less stable" (Law 2004, 6). Law did not propose a relativist vision of qualitative social research methods, but rather suggested that different practices will inevitably produce different study objects, different things, and that these objects shouldn't necessarily cohere or be reproducible. The findings that qualitative methods, especially ethnographic ones, produce are complex, messy, and fragile. This is true in the case of my own attempts at understanding and describing the complexity and the multiplicity of health and credit data. As one data marketer told me, all data can be health data.

I take this approach to underline the situatedness of sociological knowledge, as well as my own positionality and my relationships to social phenomena, and to emphasize the resemblance of data to a black box. Data's promise of total knowledge is dangerous and false. Much like sociological research, data produce their own reflexive, translational objects (Star and Griesemer 1989). So deeply embedded are data, and their uses by unseen others, in our day-to-day lives, we can never fully understand the processes by which data are harvested, decontextualized, and deployed to become the interpretable anchors for evidence or definitive factors that can predict the future. I have found that many of the practitioners, data brokers and analysts, themselves do not understand these transformations. Many of the experts that I interviewed for this book don't fully comprehend the density or breadth of the networks over which data traverse, nor can they break through the opacity of the algorithms and models deployed to interpret data. So how could I, as an outsider, do better than the professionals who work with health and credit data day in and day out? We are all insiders and we are all subjects, including the informaticians and practitioners I observed and interviewed, in the data-based society.

In the following chapters, I examine the equally disconcerting trends of the private sector's wholesale collection of consumer and financial data from a variety of sources, and the selling of this segmented information as health data to third parties. Over the years this research has taken me to hospitals, healthcare facilities, and private clinics where I interviewed medical practitioners and informatics researchers who handle patient data on a daily basis. I spoke to these groups to understand how private health data moves from the doctor's office into other sectors outside of medicine—with or without our consent. Once we are dispossessed of our data, we have little control over where they travel, how they are used, and whether our data will be used to grant us access to or deny us healthcare.

In these six years of conducting ethnographic, qualitative research, I have attended database marketing conferences and trade shows to speak with database marketers, data brokers, and sector analysts. Collecting information from these subjects helped me better understand how health and financial data become assets that are used to generate value for the credit-reporting and healthcare sectors. I have participated in patient privacy summits and academic symposia where I met with legal experts on privacy and clinical data. I have interviewed financial professionals who handle data commodities, asking them how health data is collected from credit card, retail banking, and other financial transactions. I entered all these field sites to learn more about how those working in healthcare generally, and in health informatics specifically, conceive of data as both the subject and product of profit-driven medicine.

Much of *Afterlives of Data* is the result of recent ethnographic fieldwork conducted at a university-based medical school's health informatics research institute; the fieldwork for my previous book also informs some of the research described here. The recent portion of the fieldwork occurred over a three-year period, starting in 2017 and going through the early days of the coronavirus pandemic in spring 2020. During my summers and semester breaks from the university where I teach in Philadelphia, I traveled to the medical school and hospital system, based in a Midwestern city, to learn more about how the data of tens of thousands of patients that the hospital serves are collected and operationalized into use cases, predictive scores, machine learning, and other algorithmic models in order to rationalize or to prove the efficacy of care, among many other purposes. My ethnographic fieldwork was doubly nested within a population health data research laboratory led by an epidemiologist and data scientist, whom I call Theresa. It was mostly in Theresa's lab where I learned more about how patient clinical information becomes part of massive databases maintained by healthcare providers and public and private insurers, or payers—often called data "lakes" and "oceans" by health informaticians. These databases are then trawled to make models and predictive data-based instruments designed to improve health outcomes for patients or, just as often, to create efficiencies for hospital systems under the pressures of value-based care.

Throughout this book, I have used pseudonyms to refer to the people that I interviewed or observed since commencing research in 2014. I have used pseudonyms, as well, for the institutions that I spent time in conducting fieldwork. The exceptions are the large data-marketing and health data conventions, such as Academy Health's Health Datapalooza or the Data Marketing Association's meetings and conventions, which retained their actual names. I refrain from using the name of the city where the university hospital system is located, other than to place it regionally in the Midwest of the United States.

CHAPTER OUTLINE

In *Afterlives of Data*, I examine various interests at different nodes within the larger data economy, especially those working in the credit information and healthcare sectors. These sectors collect and categorize our data to make productive data-based instruments out of our health and our debt. Many of these behind-the-scenes operations—credit bureaus such as Experian plc and insurance companies such as United Healthcare—handle information that has to do with personal finances and debt. My interest here is in how both our data and our debt intersect through our health and our lives, and how both are felt in our bodies, how we are materially subject to both. This inquiry extends to how data scientists and data advocates, in an effort to be good stewards of other people's data and in order to build trust in the potential for data to do "good" and to do justice, must navigate the subjects' histories of trauma and abuse. I examine how our data makes us both visible and legible, and toward what ends. These hidden interests have the power over our data, with little oversight and virtually no accountability, the power to categorize, define, and shape our futures. Since power demands total transparency of its subjects while remaining opaque itself, in the following chapters my aim is to reveal these processes in order to trace the afterlives that our data collectively live (Glissant 1997; Graeber 2015).

In the succeeding chapters, I reveal nodes within the network. I analyze parts, such as specific companies and informatics laboratories, and certain pieces of legislation or concepts concerning data collection or privacy—but it is impossible to provide a thorough sociological explanation of the network's entirety. And that is the point. This is how the system that trades in our data, that silently intrudes into our private lives to profit, exerts hegemony. To continue, it needs to remain all-encompassing, in the background, and utterly inexplicable. Even after all the public scrutiny of Big Tech, in the wake of the Cambridge Analytica scandal and of social media's ongoing use of personal data to drive misinformation around elections and public health, and despite widespread outcry, the needle has not shifted much. The power over our data remains with them.

The ongoing pandemic has, at least in part, shaped and complicated my thinking about data. As I was winding down my fieldwork in the population health informatics lab in spring 2020, they were pivoting their work from case-use studies and bench-level research to trying to use data to save lives during the COVID-19 public health crisis. In the response to the pandemic, access to and use of private health data became central to preventing illness and death, but access and use also became even more highly contested and controversial. Whether it was the dearth of essential demographic data on who COVID-19 was infecting or killing, or the political struggles over misinformation or manipulation of Centers for Disease Control (CDC) data, or the lifting of some privacy restrictions on sharing patient data—especially for telemedicine and for coronavirus infection contact-tracing—data in the ongoing pandemic has played both life-saving and life-threatening roles. While I remain very critical of the powerful commercial interests in the credit and healthcare sectors that access and profit from personal data about our health and our finances, especially those that continue to profiteer from the pandemic, I balance this critique with sustained consideration for those advocates who are fighting for responsible data stewardship and using data for social justice.

In chapter 1 I set the stage for the rest of the book by examining some of the ways that personal financial and health data shape our experiences in the United States' data-based, debt-based society. Some of the themes this opening chapter raises, including how data are collected from our day-to-day lives and how they are innovated upon and used to shape our health behaviors, are themes that resonate throughout the book. Chapter 2 emerged from my time spent observing a group of public health researchers, data scientists, and public-sector social workers as they built a new cross-sector data-sharing consortium. Through their efforts to share data to "do good," I explore some of the roots of the divisive mistrust in medicine and health data, and the ongoing reparative work some are doing toward making data more just, equitable, and fair. In chapter 3 I investigate the collections—the ponds, lakes, and oceans—of clinical and billing data that are produced, maintained, and reused in medicine and healthcare, and stripped of these contexts in order to be profitable data assets outside of healthcare. While these collections are regulated more tightly than other types of consumer data, Big Tech's legislative lobbying efforts have allowed large data companies outside of healthcare to access and profit from privacy-protected health data. In chapter 4 I pursue the regulatory implications of the commercial trade of debt and health data and, in particular, the market innovations around what are called "alternative data" in the financial, healthcare, and public health sectors. Chapter 5 explores how predictive models and risk scores determine or deny healthcare to patients, and how we are all scored throughout our lives. In particular, in this chapter I consider how these models and predictive scores, which are based on biased data, can replicate systemic inequities in healthcare, and worse, how they can reproduce harm even when intended to prevent it.

How we are "seen" in the data that is collected about our debt is the focus of chapter 6. Simone Browne, in her phenomenal book *Dark Matters: On the Surveillance of Blackness* (2015), analyzed how data-driven technologies, such as artificial intelligence deployed in biometrics systems, racialize underrepresented and oppressed groups, making them both invisible and "seen" by reinforcing harmful stereotypes. She connected, for example, the horrific practice of branding human beings during the American era of slavery to the modern uses of facial recognition software that contribute to racial profiling. Similarly, in chapter 6 I demonstrate that the stigmatizing effects of how we appear in financial and health data brand us and become part of who we are. Whether the "data gaze" accurately reflects who we are or how we live our lives doesn't really matter. In the end, our data are meant to make money for others. I tell the stories of several people who have struggled with how they "look" in data, how debt marks and stigmatizes them, how they struggled to become visible or remain opaque through their financial information. Like those subjected to the execution apparatus depicted in Franz Kafka's *In the Penal Colony* (1941), we, the indebted, have our credit scores or the extent of our riskiness "inscribed with rapidly vibrating and sharp needles on our skin, boring down over and over again, until the flesh falls from our bones" (Ebeling 2016, 17). All that remains are our data brands, the stigmas of our traumas made visible in data.

While Ana was alive, because of her precarity—both financial and medical—her data were marked with her health traumas and would have defined her as a "risk," an expense and a drag on the system. But in death, her data live on to produce value for myriad interests, from the hospital that treated her using her data in their own research, to Experian merging her data into an amalgam with the data of hundreds of millions of others to make new data assets that predict and score millions more. Ana's data survived her death and live myriad afterlives in all sorts of data oceans. In some ways, it brings me some comfort to imagine that Ana's data may have been used to save lives or improve outcomes for cancer patients. But the likelihood is that much of Ana's data became the data assets of third parties that had no direct contact with those trying to save her life. Ana's data have gone on to live multiple afterlives, in places identical to the ones where my data baby lived. *Afterlives of Data* results from these data rebirths, and from of all the afterlives that live on in the data about us.

1 Tracing Life through Data

> The guide explains how it's programmed to transmit biosensory information, like heart-rate, medical needs, sleep patterns. "It will be your guardian, protector. It will bring good things to you." . . .
>
> In an earlier moment in her life she might have described him as a ghost, a spiritual manifestation of the past. But she knows better now; invisibility is a prison. "Haunting" is a quaint and faint manifestation of the tortured. . . . She knew there was just one way forward and she understood the cost: the facts of her interior, available for use in a public dataset, as part of some kind of game. Besides, she hadn't made a fuss when he underwent his own erasure.
>
> "Yes, I am a pawn. Can we please go now?"
>
> —Jena Osman, *Motion Studies*

In the first few months of 2020, as the coronavirus pandemic spread across the United States, the data analytics firm Palantir Technologies won a contract with the US Department of Health and Human Services (HHS) to provide a COVID-19 contact-tracing platform called HHS Protect. Since its founding in 2003 by Silicon Valley libertarian entrepreneurs, Palantir has burnished its infamous reputation by developing platforms for military, police, and antiterrorism applications, such as Project Maven, which utilizes artificial intelligence in military drones. Palantir holds multiple contracts with the likes of the Pentagon, the Department of Homeland Security, and the Federal Bureau of Investigation (FBI). The company describes its core business as building software to support data-driven decision-making and operations. Alex Karp, founder and CEO, has characterized its platforms as clandestine services, where company products are "used on occasion to kill people.... If you're looking for a terrorist in the world now you're probably using our government product and you're probably doing the operation that takes out the person in another product we build."[1] The data-mining company is probably best known for its work for Immigration and Customs Enforcement (ICE) to develop the controversial Investigative Case Management system. Palantir's forty-one million dollar contract for its data surveillance technology has enabled ICE to intensify its detention raids on targeted communities, with serious consequences. The stepped-up raids have terrorized migrants and accelerated family separations; they increased undocumented immigrant deportations by tenfold in 2018 alone (Mijente 2019).[2] When the news broke amid the pandemic that the company had secured the HHS COVID-19 contact-tracing contract, many, especially those in the human rights and immigrants' rights communities, voiced concerns about the data privacy and security protections on the platform.[3] What data would Palantir collect, who exactly would have access to them, could the data be shared, and with whom? Could ICE use HHS Protect to come after undocumented migrants, also among the groups most at risk of contracting the coronavirus?

In a public-relations bid to assuage these concerns for a general audience, Alex Karp agreed to be interviewed by Axios journalist Mike Allen for its cable news program.[4] In the online interview, Karp, who was quarantining in his rural New Hampshire home, preempted Allen's questions concerning the public's fears about Palantir's work on a coronavirus contact-tracing tool. Karp focused on what he knew ordinary people were afraid of and what they would want to know:

> Where do you get the data? How long do you keep it? Where is it kept? Can it be removed? Is it being used to monetize me? How can I [be] guaranteed this not being repurposed for people to understand my personal life, my political views, my work habits outside of my work environment?

Allen responded, somewhat ironically, that Karp had done his job as a journalist for him. Karp's rhetorical interview questions also touch on the concerns that recur throughout this book. Who is collecting data about us? How and why do they collect our data, and who do they share it with? How is personal data used to make money for others? What stories do data tell about us? Why do we trust these data narratives, especially when these become the "truths" that increasingly define us? How and when did data become the facts of our lives?

TRACING THE INVISIBLE

This chapter opened with an epigraph from poet Jena Osman's *Motion Studies*, an essay poem comprised of several concurrent narratives in conversation with one another that revolve around visibility, the desire to disappear, and the impossibility of escape once captured and categorized by the machines of scientific inquiry and the political economy built around digital data. The first narrative is a speculative story featuring two characters, a woman and a man, who have won, by lottery, the right to be forgotten, to jump off the digital grid and "disappear beyond the company's horizon" (2019, 19). But much as, in Roald Dahl's *Charlie and the Chocolate Factory* (1964), the golden ticket wins Charlie only the chance to prove his worthiness to Willy Wonka, the lottery ticket offers Osman's protagonists only the right to run a competitive race through the desert toward digital oblivion, against other "lucky" lottery winners. The race takes the contestants ever-closer to the infinite horizon of anonymity, and in exchange, as they run, they're tracked: their movements mapped out, and their heart rates, breath, and other biometric data collected, analyzed, shared, and stored for eternity. Before they can cash in their lottery ticket, both must undergo a procedure that leaves the woman with a physical body, visible and opaque, and her partner, the man, with a body as solid as the woman's, yet transparent and invisible to human eyes (but legible to computer vision). The woman—in her opacity—realizes that even in their attempt to jump off the grid of visibility, her partner's transparent body "lives more as a trace, a clue, data" but that both bodies serve as a dialectical contrast to the other, making them both legible to the corporation's gaze (2019, 17).

The poem's second entwined narrative concerns Étienne-Jules Marey, a nineteenth-century French inventor and early photography experimentalist who was obsessed with how to make visible the body's inner, invisible movements, like the heart pumping blood through the body's complex network of arteries and veins. Marey made the body's invisible movements graphically legible through a sphygmograph, a machine he invented that traces the pulse onto a piece of paper. All of Marey's inventions visualized the unseen into graphic traces, images, and lines. These inventions also included cameras, early cinematic prototypes, and chronophotography, which sequentially captured the movement of air and fluids, the flight of birds, and the galloping of horses. Some of Marey's experiments that attempted to capture as visual information the life-sustaining movements and processes that occur inside bodies became the basis of technologies used today in seemingly disparate contexts of institutionalized power. One such technology is the sphygmomanometer, which measures blood pressure. Another device that owes a lot to Marey is the polygraph machine, or the "lie detector test." Law enforcement and other security fields in the United States still administer this test to job applicants, although the test is no longer admissible as evidence in court, as the test's claim that a subject's change in heart rate or skin conductivity indicates a falsehood has been debunked.[5] Yet as historians Lillian Daston and Peter Galison noted, Marey argued that to use mechanically generated depictions of phenomena was to speak in the "language of the phenomena themselves," and through the removal of the human, or at least of human intervention, the tendency is to let the machine take over and enable nature to speak for itself (1992, 81).

WHAT COUNTS AS DATA, WHAT DATA BECOME FACT?

Wittgenstein observed that "the world is the totality of facts, not of things" (2010, 25). How is something that is unquantifiable made to count, and to count as "fact"? When asked what constitutes data, many data scientists respond that data are information that help to anchor facts or to get to "the truth" (Leonelli 2016a). Data are abstracted qualitative information that are objectified as quantifiable values, which are then used as evidence of a phenomenon or process. It is in this way that data are made into facts that are used to help ground or reveal a truth about a phenomenon. But this of course is all contingent, shaped by the sociopolitical contexts of where and how data are extracted, who is building the models and algorithms, and how the data will be put to use.

Mary Poovey, a cultural and economic historian, writes about how the modern fact rooted in numbers and measures—data—was born in the early Renaissance. The modern fact's midwives were the European merchants and burgeoning capitalists tracking their wealth, profits gleaned from colonial empire-building (through expropriation and slavery) in the Americas and Asia—and keeping it separate from the Church (Poovey 1998). Poovey trains a meticulous eye on the rise of double-entry accounting as it developed in fifteenth-century Italy, adapted from Indian and Jewish traders who pioneered the method. It seems to be no accident that this form of accounting coincided with the Western powers' early colonization in the Americas as well as with the development of early Enlightenment knowledge production. Within her analysis, she shows how bringing knowledge about one's possessions, outstanding loans and debts, and transactions together into a ledger created connections that were at once both narrative and numerical. The ledger book was an early rhetorical attempt to make epistemic and factual statements about the world, separate from the authority of God or the Church. The double-entry accounting system was a way to confer social authority to numbers, to make numbers both expose truth and bear facts, even if the numbers were invented: "For late sixteenth-century readers, the balance conjured up both the scales of justice and the symmetry of God's world" (1998, 54).

By rhetorically making the numbers of the ledger book resemble or refer to the balance of God's order, rather than to witchcraft or sorcery, these early capitalists made numbers into facts. According to Poovey, this moment was the necessary societal shift in the West that gave numbers legitimacy and bestowed them with the authority to say something about the nature of things, about the "real world." It is all the more significant that these early capitalists used fictitious numbers to prove that the new accounting system was valid. In Poovey's account, the social and political legitimacy of the double-entry ledger book coincided with the rise of knowledge-making through the documentation and measurement of observable phenomena, another way of making numbers into facts. In the five hundred years since Italian merchants adapted double-entry book-keeping, we have seen scientific inquiry, revolutions and turmoil, slavery, and expropriation of labor, land, natural, and human resources all contribute to the making of the "modern" world and the construction of data as facts. But the process of objectifying phenomena into "data facts" necessarily involves power: power over the conditions of the extraction of the raw materials, and power over the collection, processing, analysis, and packaging of what becomes data.

Data might be understood as something that can be measured, counted, or defined by numbers or statistics—the temperature of the air, the pH level of a soil sample, or the number of people of a certain age who live within a neighborhood. For those who work in the data-based economy, data can be almost anything—the grainy surveillance camera image, the remaining charge on a phone's battery, or the blood glucose level measured over a three-month period—that can be captured and transformed into digital information. In other words, data can be any qualitative measure translatable into numbers. Once digitized, data can be transmitted, shared, and operationalized in many forms, among them spreadsheets, databases, and platforms. Surveillance cameras, credit card swipes, retail loyalty cards, and phone metadata all capture in a variety of ways what we presume to be our untraceable and fleeting actions—spending idle moments on a sidewalk, gazing into a storefront, walking into a pharmacy, or browsing the aisles of a store—and count them as data. Examining the commercial applications that track and capture consumer data, Shoshana Zuboff (2019) detailed the process of converting the intangible—behaviors and feelings—into finite, computer-readable data; these discrete inputs are fed into algorithmic models to either predict or drive future consumer behaviors.

That data and the algorithmic processes used to analyze them are neutral, unmediated, and unbiased statements on reality—that data are fact—is a persuasive and persistent notion, even in fields like medicine that rely on human interpretation. In the university population health informatics laboratory where I was based during fieldwork for this book, physicians would seek out Theresa, the head of the lab, to collaborate on research that utilizes artificial intelligence (AI) analytical techniques. In one case, an OBGYN wanted to use the lab's expertise in artificial intelligence systems and deep learning techniques to analyze focus group data she had collected from mothers who experienced trauma in childbirth. The practitioner had the expressed belief that such methods, because they removed the "human," would have less "bias" and the data could speak for themselves. In another case outside Theresa's population health informatics lab, emergency medicine and medical AI researchers at University of California, Berkeley turned to deep-learning algorithms to read and interpret knee x-rays in order to override a persistent medical bias in which doctors underestimate the pain experienced by patients from underserved populations, such as ethnic minorities, women, or poor people (Pierson et al. 2021). In this study, the algorithm measured underlying osteoarthritic damage to the knee joints to predict the pain severity a patient experienced, with an accuracy rate almost five times better than that of the radiologists who were interpreting the x-rays. One of the study's authors, Ziad Obermeyer, when asked in an interview about building AI models to support clinical decision-making, responded: "Do we train the algorithm to listen to the doctor, and potentially replicate decades of bias built into medical knowledge... or do we train it to listen to the patient and represent underserved patients' experiences of pain accurately and make their pain visible?"[6]

Ethicists of artificial intelligence Alexander Campolo and Kate Crawford (2020) call this dynamic "enchanted determinism," where magical thinking helps to rationalize a faith that AI is completely free of human bias. Many believe that AI uncovers the truth that data hold, rather than that data merely anchor a truth. But as Crawford notes in *Atlas of AI* (2021), nothing in AI computing is artificial or intelligent; rather, AI and the subjects (models, analyses, and so forth) it produces materially embody the biopolitical. "AI systems are not autonomous, rational or able to discern anything without extensive, computational intensive training" by humans (Crawford 2021, 8). Powerful, political-corporate interests build these data- and capital-intensive AI systems, and as such, the analytical results that AI systems produce become a registry of that power. From the lithium extracted from conflict-encumbered countries for computer batteries, to the "dirty" or inaccurate and biased data that are mined and fed into machine learning algorithms, AI is made from massive amounts of natural resources and human labor (and human misery), all of which remain invisible. Notwithstanding Obermeyer's and his coauthors' recognition that it is not enough to simply acknowledge the medical system's biases, AI models must be built to *account for* that bias as well as for the sociopolitical at the level of the technical. Accountability is far from the field's norm, and it is often actively resisted (Raji and Buolamwini 2019). After Timnit Gebru, former cohead of Google's Ethical AI team, published findings of the bias baked into many of the company's products, along with her public critiques of its hiring practices, which skewed against minorities and women, she was fired, along with colleagues who had defended her.[7] The "intelligence" in artificial intelligence is no more than a statistical logic that lacks the subtleties of human meaning-making, and "teaching" the algorithms is possible only by continuous human intervention.

In *Popular Science*, another section of Osman's poem, the narrative divides into the phrenological categories developed in the nineteenth century but that in some ways still have credence in the twenty-first, where one bump on the skull indicates mirthfulness, another crevice denotes indolence, and this protrusion suggests acquisitiveness. Similar to Marey in his desire to reduce life's movement to a flattened dimensionality so that it might be analyzed and made to "reveal its secrets," the phrenologists plotted their racist and classist ideologies of domination onto skulls, both to augur and to confirm their belief that the shape of one's skull could reveal an inherent truth of one's humanity (Merchant 1990, 2006, 2008). Throughout the poem, the separate narratives converge to trace the lumpy and knotted entanglements of visibility, data, information, and the relationships of power and control that arise when subjects are brought into the visual realm to be known, fixed, or activated by power. The data are not just left to speak for themselves, but are always filtered through the lens of power: the power to collect, interpret, and repurpose data for power's own ends. As data ethicist Louise Amoore argues in her book *Cloud Ethics* (2020), it is not simply a question of data visibility, but of how power *perceives* and *makes* subjectivities through algorithms, and how this perception constructs, shapes, makes possible, or forestalls the subjects who have been made visible. When the subject is perceived by power, she notes, "the transformation of perception involves changes... what could be brought to attention, changes in the horizon of possibility of human action" (Amoore 2020, 16).

Significantly, the visualization of the invisible, or of what's hidden under the skin, as Osman's poem suggests, becomes known as "data."

These data produce evidence and make facts through a process that captures the fleeting robustness and ruddiness of life and flattens it into a two-dimensional line or into the binary code of a yes or no value. After one's life has been translated and made perceptible as a binary object—as data—the second process, legibility, entails a phrenological approach to data analysis, a way for power to "read" you further into subjectivity. Once legible as "data," as sociologist Mauricio Lazzarato noted in *Signs and Machines* (2014), individuals are transformed into "dividuals," following the Deleuzian concept of computational control, to become "mere inputs and outputs, a point of *conjunction* or *disjunction* in the economic, social, or communicational processes run and governed" by late capital's statistical logics (Deleuze 1992; Lazzarato 2014, 26, emphasis in the original). Gilles Deleuze noted that in the beginning of the 1970s, as business was beginning to use computing more widely—an era that saw the rise of the computerized databases that power and capital leverage as a means of decentralized control—we came to understand ourselves and to be understood by others as "dividuals," or masses of algorithmically categorized individuals, who can be governed and controlled (Bouk 2015; Deleuze 1992). From these flattened lines, these glimmering points of data, a narrative of your life escapes you, only to be told as a story of estrangement.

INTERPRETING LEGIBLE DATA AS COHERENT NARRATIVES

In historian Saidiya Hartman's essay "Venus in Two Acts" (2008b), Hartman revisits the life of a girl, kidnapped and murdered in 1792, and given the name Venus by her captors—the sailors and slave traders on the slave ship *Recovery*. Hartman first encountered Venus while researching the archives of the Atlantic slave trade for her heartbreakingly beautiful book *Lose Your Mother* (2008a). In the essay, Hartman reassembles the girl's moments before her death at the hands of the ship's captain into a coherent but highly speculative narrative. Hartman uncovered a single-sentence reference to Venus and her murder buried deep in the notebook of a sailor who witnessed her torture and death. Hartman speaks to her own desire to weave a narrative that restores life, and justice, to an existence all but obliterated from the archive, save for the scraps of evidence that show she once lived. With one point of reference to Venus's life—the fact that she existed—Hartman built a narrative to restore her humanity.

While a historian's use of archival scraps to build a narrative of a life may appear to differ greatly from the interpretative work of analysts handling the data of millions of strangers, it does not. With just a few clues, analysts categorize, interpret, and make personal data conform to what are often predetermined narratives. For both the historian and the data analyst, the lens of power filters life stories. But this is where the work of a historian like Hartman and that of a data analyst working in commercial settings diverge. For Hartman, restoring visibility, life, and dignity to those erased from the archive is a form of justice, whereas Big Tech data analytics is an expropriation of life for profit.

In our own time, the cruel and stark power that excavates and narrates data in order to subjugate was laid bare when a whistleblower exposed to public scrutiny a US government spreadsheet, maintained in secret, about detained pregnant girls and young women.[8] The anonymous leaker revealed that Office of Refugee Resettlement (ORR) (an office housed within the HHS) officials had maintained a twenty-eight-page spreadsheet listing every pregnant teenager and young woman seeking asylum and held in detention camps at the US-Mexico border. The spreadsheet's columns included the age (with some as young as twelve years old), when a pregnancy test was administered and if it confirmed the pregnancy, the estimated gestation age, and if the pregnancy was the result of rape or consensual sex. The "Notes" column detailed in the dehumanizing language of acronyms the excruciating trauma that the detainees had survived on their dangerous journeys to the US border seeking refuge and safety. Such information included the ages of the men that had assaulted them or, in cases of consensual sex, the ages of their boyfriends, as well as more contextual details in regard to assaults (for example, "Reported to be raped by uncle in COO [country of origin]"; "consensual relations however UAC [unaccompanied alien child] reported rape by unknown assailant @ 1 month into pregnancy").[9] The ORR officials maintained the spreadsheet on the detained girls and women at the same time that the HHS had "lost track" of the more than 1,480 children that the Department of Homeland Security had separated from, and was later ordered to reunite with, their families.[10] Human rights organizations such as the American Civil Liberties Union speculated that the spreadsheet's purpose—its collection of so much fine-grained data on one group of detained children while other children were literally lost—was to prevent pregnant girls and young women from obtaining abortions and to maintain control over their bodies at the level of cells inside a uterus.[11]

HOW LIVELY DATA DEFINE US

Osman's speculative prose poem plucks at the gossamer threads that trace our lives, our pleasures and desires as well as our traumas, and connects them to the larger and unseen systems that categorize and define us as consumers, as debtors, as patients, as criminals, as a cluster of behaviors that can be predicted—as data. We may be categorized as citizens, say, by the likes of political election pollsters, only in the service of weaponizing our data and dividing the electorate. Osman's poem underlines how these digital systems, and the political economies that make them possible, have an insatiable appetite for our data, rendering us as both the raw material and the product. In my previous work *Healthcare and Big Data*, I described the results of the transformation of personal data into novel and unrecognizable data-based products that are commodified and sold as "lively data" (Ebeling 2016, 95–132). I attempted to retrace how information concerning my health, and the debt I incurred to fund my medical care, had escaped both the clinic and my financial records, only to end up in the databases of marketers and data brokers selling consumer data. Through this hunt, I came to see this purloined data as the flesh and bones of my dead baby, reanimated by marketing midwives and reborn as "lively data." In the case of my miscarriage, my lively data was made to conform to a constructed marketing narrative—a healthy, full-term baby that needed me to buy it things—that was completely ignorant of or ignored my heartbreak so that corporations could profit. For others, lively data could be a cancer diagnosis, or a flood that destroyed their home, or a lost job and a consequent loss of health insurance. Lively data are the reborn fictions that conform to the trite categorizations and narratives built for us by data analysts and brokers, marketers, and others in Big Tech or elsewhere in late capital who seek to profit from our lives.

Throughout the process of having the viscera of our lives fed to the machines, we are not considered agents with free will or a choice in the matter, despite the empowerment rhetoric from Big Tech that encourages us to "take control of your data, take control of your life." Any choice we think we make has been shaped a priori by a marketing team that developed the fine-grained segmentations that categorize consumers or by software engineers who coded data to fit into a model or an algorithm.

Once these systems define our data, and thus define us, it is very hard to change how the systems classify the data or make them adhere to narratives. Cultural and communication theorist John Cheney-Lippold dubs these definitions our "algorithmic identities," in which we are "understood in the datafied terms of dynamic, soft-coded, and modulating measurable types . . . that suggest a new relationship to power," or what he calls "soft biopolitics" (Cheney-Lippold 2017, 36). For some, data definitions can be mildly annoying, but for others, they are deeply biased and harmful. The harm is often baked right into the datasets used to teach machine-learning algorithms. Data that are inaccurate, false, misleading, or shaped by biased assumptions can make the task of building accurate models difficult. While there may be some efforts to scrub data for bias before datasets are entered into algorithmic models, millions of commercial as well as public or "open" datasets remain "dirty," and they continue to produce harm.

The categorization and labeling of individual photos within the Massachusetts Institute of Technology's (MIT's) 80 Million Tiny Images training library is just one of many examples of how dirty data can produce harm. In 2008 MIT computer engineers compiled millions of digital photos of things like stoplights or appliances, as well as people, animals, and plants, in order to train algorithms to detect and distinguish between objects in digital images. Each photo in the massive collection comes with an image description label in words short enough for computer vision algorithms to "read." The training dataset has been used to benchmark computer-vision algorithms and to teach all sorts of models used in AI applications, ranging from facial recognition software to radiological diagnostics. The problem with the 80 Million Tiny Images is that the library was never vetted for how it labeled the images depicting people. The MIT engineers developed the library from more than 79 million images scraped from Google Images, which they then arranged into seventy-five thousand categories.[12]

Images depicting people constitute about 23 percent of the dataset. When independent researchers validated the dataset in 2020, they discovered that it individually put into classes (the 75,000 categories) or labeled with misogynistic and racist language many of the photos of women or people of color, and that it used some images to represent demeaning stereotypes (Prabhu and Birhane 2020).

Facial recognition technologies driven by artificial intelligence fed with poor-quality or "dirty" data, like 80 Million Tiny Images, go on to target, profile, and usually misidentify Black or Asian faces, at a rate of between ten and one hundred times more than white male faces (Browne 2015; Garvie, Bedoya, and Frankle 2016; Noble 2018).[13] Michigan state police signed a $5.5 million contract to a privately held company called DataWorks that fed millions of low-quality images sourced from disparate databases of drivers' licenses photographs and surveillance video stills into the platform's machine-learning algorithm. The software falsely identified Robert Julian-Borchak Williams, a Black resident of Detroit, as a shoplifting suspect, and the police arrested and charged him with larceny, despite his innocence.[14] Facial recognition software serves as one example of how, when biased data are used to teach machine-learning algorithms, they become the fictions that reinforce social falsehoods. Former Google AI ethicist Timnit Gebru, among many others, such as her co-authors Joy Buolamwini and Deborah Raji, has long argued for transparency and the contextualization of datasets that are used in the machine learning deployed in decision-making in the high-stakes domains of healthcare, policing, and education (Buolamwini and Gebru 2018; Gebru et al. 2020; Mitchell et al. 2019; Raji et al. 2020).[15] But these algorithmic models tend to be useful only to those who build them. When models operate as idealized fictions of the truth, the real harm is when these falsehoods harden into facts (Appiah 2017; Peterson 2021).

FACTS OF LIFE IN DEBT DATA

The late-capital formation of socioeconomic and political power over and through the capturing and monetizing of personal data that are constructed as the "facts" of individuals' lives has been characterized as surveillance capitalism (Poovey 2018; Zuboff 2019), data colonialism (Couldry and Mejias 2019; Couldry and Yu 2018), a black box society (Pasquale 2015), and through the lens of moral economies and debt societies (Fourcade 2017; Fourcade and Healy 2013; Graeber 2015, 2014; Lazzarato 2014, 2015). The core of these frameworks conceptualizes a political economy that adheres to the notion that personal data is a valuable commodity, extracted from consumers' lives and bodies by powerful commercial interests, like the Big Tech giants Alphabet or Amazon. These corporations dominate the information technologies and digital, online services sectors. Once commercial interests capture the data as "raw material," they innovate upon the data through a variety of processes. Such processes include the de-identification procedures necessary for data owners to make data compliant with privacy and security regulations, cleaning and categorization methods that are used to manage data, and algorithmic analytical practices that construct models or platforms to "make sense" of data. All these processes (and more) result in novel data-based products that can be sold onwards to third parties as data commodities. In this way, data become valuable financial assets that generate profit for the innovators, and that are completely estranged from the original source of data, the individual. In a way, our data are more than the assets of corporate power: They are our collateral in the debt-based society.

Debt of all kinds—from consumer to medical—and at all levels—from personal to national—governs American society. Social theorist Mauricio Lazzarato has dubbed this distinctly American cultural phenomenon the "debt society" (Lazzarato 2015, 61). A society that debt organizes and governs is one where financial and speculative capitalism hegemonically exerts the debtor-creditor relationship onto all social interactions, and solidifies a debtor's subjugation to the world-ordering power of late capital. Debtors are ensnared in an asymmetric and dialectical relational trap with those who give credit. A debt society is one where the terms of democratic power are turned upside down: workers can't strike, citizens can't protest, even consumers can't boycott because their lives are yoked to the debt they carry. For example, workers who depend upon employer-sponsored health insurance for their medical care—about 55 percent of Americans are covered by insurance they get from their jobs—are less likely to organize for or protest the conditions of their employment for fear of medical debt (Keisler-Starkey and Bunch 2020). When citizens rise up collectively and march to demand social justice and change—like the widespread 2020 Black Lives Matter protests that mobilized millions to the streets—they risk being exposed to state violence and subjected to police arrest or injury. For those protestors that put their bodies at risk in demonstrations, they must overcome the fear of debt (among other fears, certainly) generated by the carceral state—from legal fees and bail bonds to the debt generated through imprisonment, as well as the threat of wages lost while incarcerated. Consumers already in debt to a corporation, through retail financing or credit card debt, have no way of "voting with their wallets" when their wallets are the collateral that the corporation holds.

The origins of a debt security—the bond—serve as a poignant reminder of what is at stake, and of the biopolitical bedrock of the debt society. A bond is a financial instrument that serves as a loan to any entity—be it a corporation or a city, state, or national government—that issues the note. An investor who buys stocks becomes the owner of a stake in a publicly traded company; if they buy a bond, they become a creditor who gives a loan with a fixed rate of interest, with both the principal and interest paid back at a future date. The bond, at least in the American context, is rooted in slavery. Some of the earliest bonds issued in the United States mortgaged enslaved people: the creditors giving loans considered slaves a more valuable, liquid, and flexible collateral than land or real estate (Murphy 2013, 2021; Wells 2020).[16]

The contemporary American debt society is haunted by the brutality of its slavocracy origins: at the core of debt is a human life, abstracted as a collateralized instrument, that returns a promised value to those who possess it. This legacy is obvious in one of Wall Street's latest innovations: the social impact bond. The social impact bond is an investment instrument that leverages, say, the health outcomes of welfare recipients or parolees' recidivism, in exchange for a loan, in the form of a bond, to a social services organization (Olson and Phillips 2013; Rowe and Stephenson 2016). When corporate interests turn to the financial, health, and other personal information of millions of debtors, the promised profits and wealth drawn from collateralized life are enormous. While public-facing Big Tech corporations may attract more public attention and ire about how they profit from our data, it is those powerful interests that go unnoticed, yet are essential to the debt society's functioning, that realize the true value of our collateralized lives.

Often the usual suspects of Big Tech—Alphabet and Google, Facebook, and Amazon—tend to be the focus of public pushback or resignation to the fact that these firms vacuum up some of our most intimate data, information about our health and our debt. Yet because the credit reporting and financial information services sector primarily conducts business-to-business operations, it attracts little public scrutiny over how it profits from personal data about debt. And the profits are huge. For instance, companies such as Experian plc and Equifax, two of the most powerful personal credit bureaus in the United States, assetize and profit from the personal and consumer data that they collect on more than 98 percent of American households.

How did the extraction, production, and collection of information about our lives become such precious commodities, ones that underpin virtually every sector of the American economy? Companies working in this part of the credit services industry enter into complex partnerships in the public and private sectors to access and profit from personal data. For example, MasterCard Incorporated, the multinational financial services company behind the MasterCard credit card brand, partners with state and federal social benefits programs, such as the Supplemental Nutrition Assistance Program (SNAP), to provide payment and transaction services through branded debit cards distributed to benefit recipients (Hahn et al. 2020).[17] Such partnerships collect the transactional consumer data of millions of Americans, which become these companies' data holdings and can be shared onwards with third parties. Virtually all credit and debit card brands, like MasterCard or Visa, have data-sharing agreements with credit-reporting companies like Experian.

In his history of the American credit rating industry, *Creditworthy* (2017), Josh Lauer connects the rise of the burgeoning "information economy" (the commodification of and trade in data) to changes in capital formation, which transformed from industrial to speculative, financial capitalism starting in the mid-nineteenth century and continued into the twentieth and twenty-first centuries. When the mercantile agency system, or what is now called credit reporting, began, Lauer notes:

> The idea of an information economy—one in which knowledge might be packaged and sold as a commodity—was apparently inconceivable.... The mercantile agency system of the 1840s introduced an entirely new way of identifying, classifying, and valuing individuals as economic subjects.... [What was] invented was not just a highly coordinated system of disciplinary surveillance, but the very idea of financial identity itself. This new technology of identification became a key infrastructural component of the modern credit economy and, in turn, produced its own category of social reality. (Lauer 2017, 49)

Through freelance as well as in-house reporters embedded locally in the towns and villages, early credit bureaus collected and collated intelligence, at times incredibly personal and sensitive information, on individuals. These reporters would document various types of data, all qualitatively understood, based on vague character reports collected from informants who personally knew the target of intelligence gathering: Was the person of "good moral standing" in the community? Was the individual a regular churchgoer or was the person known to be "honest" in their marriage? As the industry matured, it earned a reputation of peddling nothing more than hearsay. Those in the business of producing material commodities—from steel to soybeans—viewed the mercantile agency system as nothing more than "parasitic middlemen" dabbling in immaterial goods that amounted to nothing more than gossip (Lauer 2017, 49).

To burnish this collective bad reputation, the industry on the whole attempted to "quantify" the qualitative information that it held, by including financial records and information held by creditors, too. The principal techniques of acquiring this information were compelling banks to share records and forcing businesses to share balance sheets and other financial reports. The agencies also employed full-time, in-house credit reporters who collected and updated such information (Lauer 2017, 40). By the mid-twentieth century, the information files that credit bureaus held on increasing numbers of Americans included each customer's name, occupation, income, marital status, age, address, length of occupancy, bank accounts, and even the person's physical appearance and other personal details (114). The industry also tied one's existing wealth to the credit rating, so that reputation or qualities that are harder to quantify no longer had as much significance in assessing creditworthiness. A person's identity was comprised of data, which "came to replace the living person as an embodied index of trustworthiness" (113). Agencies also implemented a numerical rating system, to provide the data with the sheen of unbiased, objective information. One agency, Dun, implemented a system that ranged from A1 to 4, with the top of the scale, A1, denoting the highest rating, worthy of unlimited credit access. Any business or individual that had a capitalization of more than one million dollars automatically earned an A1+ rating, regardless of the firm or person's "moral character." This numerical rating system, where the diversity of one's data held by credit bureaus is condensed to an "objective" number that supersedes and brands one's character, has become its own social reality: the credit score. Data about personal debt and credit scores have become the data-based "facts" of an individual's life. These data facts of our mortgaged lives circulate ever more widely as commodities that return value to those who stand to profit from the indebted (Ebeling 2016).

Like Marey's machines, which made both invisible processes and the immaterial legible, in the following chapters, I expose some of the ephemeral afterlives of data. I trace how health data and personal financial information circulate and are reborn in predictive models or as scores used in high-stakes decisions about one's healthcare or access to credit. These data-driven processes and algorithmic solutions are so institutionally normalized that they have become the facts that shape and determine life chances and outcomes for millions.

After less than a decade of the data economy's prominence, it has become a cliché that data are the commodity that runs the economies of late capital, and that the assetization of data determines the lives of many. Once life is made legible in data, it presents profit opportunities for Big Tech and for the thousands of unseen others that feed off the data value chain. The quantification of everyday life subsumes virtually all interactions—from how many "likes" an online post may get, to how many steps one takes a day, to the highs or lows of a credit score. In the data-based debt society, everything is made to count, and everything is made to return financial value. It must, in order to make human life conform to the statistical logic of late capital.

2 Building Trust Where Data Divides

Our first meeting is in a low-rise, nondescript brick building on the edge of the sprawling hospital and medical school campus. Compared to the surrounding glass and steel towers, this is one of the older buildings on the block. The campus has undergone a construction boom within the last decade, accompanied by neighborhood redevelopment projects that depend on a city's existing universities and hospitals as a focus for expansion and revitalization, or the "meds and eds" model of urban renewal. This model attracts the "creative class" to dying city centers (Florida 2014). The campus is little more than a mile south of Division Street, a major corridor that splits the city in half, physically and symbolically, between north and south. This Midwestern city divides along racial and economic fissures as do many American cities, and it continues to be deeply segregated and unequal. Neighborhoods to the north of Division Street have a majority of Black residents and endure the legacies of forced segregation, redlining, environmental racism, and divestment. Neighborhoods to the south of Division Street are whiter and comparatively more affluent, and historically they have benefitted from investment in infrastructure, schools, and other community resources. The Division, as it is known, represents the historic and ongoing inequalities along racial and class lines that continue to divide the city.[1]

We are gathered around the conference table to talk about sharing data. Our focus is not just on any data, but on the information hospitals, regional planning offices, public health facilities, police departments, and social service nonprofits collect. A lot of the data—patient health records, arrest records, public health surveillance information on the rate of syphilis infection within a zip code, say—are very sensitive and can be stigmatizing if linked to an individual. Only a handful of people have organized this data-sharing initiative, but the organization's aims are broad and somewhat daunting: to rally disparate social service organizations and local government departments—all with their own contending interests and histories of mistrust among them—to share their data resources in a centralized hub. The group gathered around the table hopes that sharing data will improve the lives and communal well-being of residents in this Midwestern city.

THE DATA-SHARING CONSORTIUM

Though this effort to build a regional data consortium (the Consortium) is still in its infancy, the initiative has support from two of the most influential universities in the city. And all of us in the room, including myself, are academics of various stripes. Steve, who has a degree in social work, directs a community partnership for the public university that launched the Consortium. A few years earlier a local business magazine recognized his work in social and community innovation by naming him one of the "'30 under 30' to Watch." In turn, Emma is a researcher in public health surveillance with a specialty in sexually transmitted diseases, and Theresa is an epidemiologist and data scientist focused on cardiovascular disease. Emma and Theresa are colleagues at the private university that is hosting the morning's meeting. Their university, Central Midwest University (CMU, a pseudonym), boasts one of the strongest medical schools in the United States.[2] CMU's medical school is affiliated with the largest hospital system in the Midwest, which I will call CM Healthcare. The CMU medical school and the hospital system closely collaborate on medical research and training. On the technical side, the Consortium recently hired Carl, an experienced data manager coming from the nonprofit and health advocacy world. He is the only team member with experience outside of higher education. Then there is me. As a visiting researcher whose work generally steers clear of statistical analysis or quantification, I am humbled by what I don't know. I am here to observe and to learn.

The CMU and CM Healthcare campuses are situated in the center of a metropolitan area of more than eight thousand square miles, which sprawls from a depopulating city to suburban counties (including some of the country's wealthiest) and rural communities. The city has one of the highest poverty rates in the United States, one which cleaves along racial lines, with Black residents experiencing higher rates of poverty than white residents. Historic and systemic factors drive the region's endemic inequalities. As a result, many metro-area residents contend with ongoing isolation, limited healthcare facilities, inadequate infrastructure, and other insufficient essential services, like unaffordable housing and unreliable public transportation, that directly impact their health outcomes (Gordon 2008). Both CMU and CM Healthcare consider it their mandate to serve all of the communities situated within the entire metro area, and the collective hope for both institutions is to reverse some of these entrenched problems that residents face. While success has been mixed, those around the table see the fledgling Consortium's use of data to bridge regional inequities as a start.

DATA VISUALIZATION AS EVIDENCE ANCHOR

Steve, Emma, and Theresa started this collaboration about six months earlier by building a data visualization tool to demonstrate the power of data sharing and to help neighborhoods and communities construct a deeper understanding of some of the health challenges they face. Steve calls this "using data for good." The prototype is a public health dashboard—a web-based, digital platform. The Consortium will use the dashboard to visualize, through graphics and maps, the shared data that the organization hopes to bring together into one central hub. The organization hired Carl to facilitate the technical relationships necessary for sharing and standardizing the disparate databases that will become the backbone of the data infrastructure.

Some early prototypes of the dashboard display a regional map with filters that can be applied as overlays onto the map. In these early stages of development, the Consortium has fleshed out only two filters: one uses regional infant mortality data, and the other uses the rates of chlamydia infection from a public health dataset. As Emma toggles through the data layers, red dots undulate across the map. Each dot represents one positive case of chlamydia infection reported to the city's department of public health, as the Centers for Disease Control (CDC) mandates. Red dots cluster in some parts of the map, while other areas are starkly blank. The dots align with the blankness of vacant lots in the north of the city, and they are nearly absent in the hustle and bustle just south of the Division on the map. Other data layers could show, say, pedestrian fatalities in a neighborhood or the rates of sexually transmitted diseases. Both issues are of concern to public health researchers because they can indicate social barriers or determinants that impact health outcomes. Pedestrian fatalities can indicate unsafe street infrastructure like a lack of sidewalks or safe crosswalks for pedestrians, while high obesity rates might be explained in part by a lack of grocery stores within walking distance for residents. In turn, the lack of a public health clinic that provides free STD screenings and treatment could be a reason why emergency rooms are reporting high STD rates in their patients. These indicators (and many others) comprise what experts consider to be the social determinants of health: access to healthcare and physical infrastructure. The Consortium hopes that the data, when constructed as visual representations, will help to portray these types of divisions as they play out in measurable health outcomes, as well as help community members to see how their data can help visualize solutions. Visual representations, it is hoped, can put flesh on the bones of the data to tell the human stories behind the numbers.

The Consortium designed the dashboard to evolve. As more stakeholders from regional public health departments and clinics share data with the Consortium, additional layers and more dots representing different social factors that impact community health will make the dashboard more robust and functional. For example, Emma hopes that stakeholders will share more information about pre- and postnatal health in the region. For population health researchers like Emma, the regional infant and maternal mortality rates are very concerning, as they have been rising not only in the city, but also across the state, and are some of the highest in the country.[3] When these already stark mortality rates are disaggregated on the race and ethnicity of patients, the data demonstrate that Black mothers and their babies experience negative outcomes four times higher than those of white mothers and their babies (Singh and Yu 2019). These neighborhood maternal and infant mortality rates serve as key indicators of the overall health of a population and point to just how unequal our healthcare system is. Emma wants to use the dashboard's data visualization as an "evidence anchor" to argue for CM Healthcare to expand access to prenatal (and other) healthcare services in areas of high health inequity (Halpern 2014; Leonelli 2016a; Tufte 2006).

Around the table, the discussion moves to data-sharing models that can serve as exemplars for the Consortium to emulate, such as the Cardiff Violence Prevention Model. In this model, policy planners in the Welsh city combined data from hospitals and police records of violent crime reports to map the public health implications of violence (Florence et al. 2011). The CDC, among other national institutions working in social and health services, has highlighted the Cardiff Model as a success story. The Cardiff Model fosters cooperation between partners and stewards in data governance to share data that empower communities and, its adopters hope, solve endemic public health problems. The CDC has encouraged public health centers to widely adopt, adapt, and implement the model to address a variety of health challenges that the communities they serve face. The Cardiff Model works only because of strong community partnerships and robust trust relationships between the data partners, Emma noted. "It's their 'secret sauce,'" she said. The Consortium is following the lead of other American cities, such as San Diego and Philadelphia, that are building their own data-sharing initiatives inspired by the Cardiff Model, in an effort to change the social conditions of the region's residents.

DATA-SHARING DIVIDES

Many of these collaborative efforts between city governments and the private sector began in the wake of an Obama Administration Office of Management and Budget (OMB) memorandum, "Open Data Policy:

Managing Information as an Asset," which outlined new requirements for all government data to be "machine readable" and accessible by the public (Burwell et al. 2013). All publicly produced datasets must be available online and standardized, as through an application programming interface (API). This online availability allows the communities from which much of the data derives to search, download, and use the information for their own purposes. Through the executive order's mandate, taxpayers' funds produce all data local governments hold, so the mandate considers the public to "own" the data. Local governments serve merely as stewards and, therefore, the data should be available online and easily accessible for the public (Ebeling 2016, 106).

Following the Open Data mandate, many municipal and regional governments, in collaboration with the private sector, created open data-sharing initiatives or community information exchanges (CIEs), to varying degrees of success (Thomas 2017). For instance, in 2009 Congress passed the Health Information Technology for Economic and Clinical Health (HITECH) Act, a culmination of more than twenty years of regional efforts across the United States to create health information exchange hubs. Through these hubs, healthcare systems could use sharable and open data to improve patient care and, it was hoped, prevent medical errors, lower delivery costs, and improve patient trust (Dullabh et al. 2011).[4] In another example, in 2015 the Robert Wood Johnson Foundation founded Data Across Sectors for Health (DASH) to promote data sharing among multiple sectors outside healthcare, such as housing, social services, education, public safety, and economic development public organizations. DASH's goal was to improve community health through ten exemplar community organizations across the United States. Later, All in for Community Health spun out of the 2019 DASH initiative and had the aim of expanding the cross-sector data-sharing model to 125 community collaboratives (O'Neil et al. 2019). The Consortium is, in part, modeling itself after these efforts to share and use data across public sectors that provide services to residents and manage the public health of a city or region.

These open data initiatives comprise part of the larger mandate in the medical sciences to share research data and fundamental knowledge, based on a widely held belief among scientists that sharing data accelerates discoveries. These appeals for more data openness, while well intentioned, may come into direct conflict with very real social pressures for scientists: career trajectory or status in a particular field, the drive toward patenting and scientific capitalism, or the quest for highly competitive and scarce research funding (Gewin 2016; Leonelli 2019, 2016b; Strasser 2019).

Because scientists have made demands for transparency in the governance of research data, many open data initiatives, such as the Consortium, are following what are known as the FAIR principles for data management: they are making research data *f*indable, *a*ccessible, *i*nteroperable, and *r*eusable (Wilkinson et al. 2016). But the FAIR principles for scientific data management may not be entirely applicable to personal data, and they are even more problematic for private health data that are individually identifiable or can be rather easily de-anonymized (re-associated with the individual). In the health and medical sciences, policymakers and the scientific community have made similar open data appeals, especially in the era of genomics and personalized medicine (Milne et al. 2019). While health informatics researchers who use patient data often call for FAIR principles to be applied in research, especially to interoperability—the ability for different computer systems and software to exchange information—patient privacy and consent are often the sticking points in attaining "frictionless" data. But two researchers sharing, say, datasets of bacteria genomic sequences is quite different than an emergency room sharing patient data with a police department (Leonelli 2016a; Strasser 2019).

While some initiatives, like the Cardiff Model that the Consortium advocates, have received praise for their positive impact on health and community safety outcomes, many more data-sharing models are fraught, seriously flawed, or downright dystopian (Eubanks 2018).[5] From China's social credit system to India's Aadhaar data platform, many data-sharing systems are built in politico-economic contexts so that governments, supported by and only possible with the technological prowess of Big Tech, can surveil, track, and control racial, ethnic, religious, and other minorities and vulnerable groups. These platforms also represent the erosion of the thin line that separates authoritarian governments from the tech titans that engineer the artificial intelligence and tracking platforms. For example, a document leaked to the *Washington Post* revealed that the tech giant Huawei had developed facial recognition software that would send government authorities an automated alert when a surveillance camera detected a "Uighur face," indicating an individual belonging to a persecuted minority group.[6] These data-sharing structures, built by authoritarian capitalism, range from the coercive to the fatal.

China's social credit system appears to be a coercive data infrastructure. Through a series of databases and initiatives that collect and monitor data on every individual, governmental institution, and corporation within China, a "trustworthiness" score is developed and maintained for each individual or organization, similar to a financial credit score. The rating system collects information on an individual from a diversity of sources, including registry and government offices, court and criminal records, and third-party sources, such as mobile phone retailers. But the system also collects "anti-social" behavioral data, such as sidewalk video surveillance to catch jaywalkers or smokers in smoke-free zones. The system aggregates the data and uses it to "score" a person. A high social score can ensure that an individual has easier access to higher quality healthcare, and a lower score can mean job loss, fewer educational opportunities, and travel restrictions (Engelmann et al. 2019). While Western reporting on China's social credit system tends to emphasize the negative and dystopian aspects of the platform, some ethnographic research investigating how ordinary Chinese citizens feel about social credit scores suggests a more nuanced and culturally specific understanding of the system, with some who were interviewed explaining that they are happy to give up some privacy for more security and certainty in their daily interactions.[7]

For some Indians, the national data-sharing systems have become deadly. In the early summer of 2019, Ramachandra Munda, a sixty-five-year-old man, starved to death in his village in the state of Jharkhand. His family alleged that coercive data sharing spurred his death. While it is true that people often starve not due to natural disaster but rather to political failure, Munda's death symbolized the specifically *digital* and political causes of starvation. To receive his weekly food subsidy from India's Public Distribution System, Munda submitted his fingerprints to a biometric scanning machine. After the scanner in Munda's village broke and despite his efforts to obtain his ration with his paper ID card instead, the local distribution office refused to disperse it to him without the requisite scan.

After four days without eating a single grain of food, as his family later told news media, Munda died.[8]

Under the Narendra Modi nationalist government's initiative Digital India, the Aadhaar scheme links each Indian's biometric identity with his or her national ID number and numerous public and private services. The Aadhaar program uses an individual's fingerprints to digitally unlock services and to enable essential activities of life. One's fingerprints can access everything from food subsidies to mobile phone services, and even allow government employees to clock in at work. More than an online trace left by internet browsing habits, the Aadhaar system quite literally tracks the bodies of the Indian citizenry. Politicians argue that the digital platform, which is officially voluntary, delivers social services more efficiently to "empower residents of India." If you are hungry or need a job or a doctor, however, registering your fingerprint in exchange for food or life-saving medicine seems pretty mandatory (Ebeling 2020, 37). The consequences can be far-reaching, especially when the system collects personal data and shares it between the public and private sectors, and within the larger context of controversial legislation. For example, the 2019 Citizen Amendment Act officially excludes Muslim and Jewish immigrants from obtaining Indian citizenship, the National Registry of Citizens requires all residents to prove their Indian citizenship, and laws restricting interfaith marriages imperil the rights and lives of vulnerable groups, especially religious minorities (Bhatia and Gajjala 2020; Mishra 2020).[9] These laws, already weaponized against Indian minorities, especially Muslims, have the potential to be more deadly when supported by digital tracking and data-sharing platforms.

In the United States the public is increasingly mistrustful of data sharing, particularly information sharing between government and the private sector (Auxier et al. 2019). With the world's highest rate of civilians murdered by law enforcement, US police departments may be willing to share crime data they collect to feed public health models like the Cardiff Model, but they do not even collect data on the number of shootings or deaths of civilians at the hands of the police.[10] Without such data, models cannot predict or prevent police violence. Ezekiel Dixon-Román, a social scientist who studies decision-making algorithms used in education and criminal justice, warns that many of these data models replicate the racialized biases that they seek to overcome (Dixon-Román 2016). Human rights advocates voice growing alarm about the increasing reliance on similar data-sharing agreements among healthcare, public health, social services, and the criminal justice systems, which can share data that feed predictive risk or automated decision platforms. These agreements often result in intrusive surveillance and punishing control of vulnerable groups. Philip Alston, the special rapporteur on extreme poverty and human rights for the United Nations, in a report on the algorithmic surveillance of poor people, identified several predictive data platforms that the United States, the Netherlands, and the United Kingdom use to administer the welfare state. He observed that the platforms were little more than expensive excuses, funded by taxpayers, for governments to "slash welfare spending, set up intrusive government surveillance systems and generate profits for private corporate interests."[11] In a press statement, Alston in particular noted problematic initiatives such as the recidivism prediction tool that Philadelphia uses to evaluate the suitability of inmate probation, and the System Risk Indication (SyRis) data surveillance that Rotterdam uses to detect benefits fraud. Alston claimed that governments sell these systems to the public as a form of "citizen empowerment," with declarations that digital solutions will deliver more efficiently administered governments and higher rates of well-being (Alston 2019).[12] Most of these initiatives, however, are public-private partnerships with powerful technology companies, such as IBM or Palantir.[13] And when there is no legislative or regulatory power to hold those who misuse data (or even just build sloppy systems) to account, or when these abuses are baked in and officially sanctioned, there is no reason for citizens to trust these initiatives. Essentially, it is the citizens themselves, especially the most vulnerable, who are delivered to the corporate surveillance state on a platter.

An epidemiological disaster, such as the COVID-19 pandemic, brings to the fore these tensions between the need to share data in an emergency and the privacy concerns of citizens and patients. Compounding these tensions during the pandemic is the partisan, ideological war over data sharing, in which state governors or local officials misrepresent infection and death rates. Rebekah Jones, a public health data scientist working for the Florida Department of Health, was fired for refusing to fabricate data on the coronavirus on a public-facing portal so that the data would support reopening the state. Jones later built her own data dashboard to share vital and accurate information directly with Floridians.[14]

The Consortium's hope is that by acting as a bridge builder between institutions that hold data and the communities that they serve, it will allow data to promote trusting relationships between public sector institutions and neighborhoods, rather than to surveil or punish communities. Through this facilitation role, the Consortium members believe that data can perform a kind of social work. As philosopher Sabine Leonelli has noted, data are relational objects that we use as evidence to anchor knowledge and truth (Leonelli 2019). The Consortium wants to use data to anchor trust. But to make data do this social work, one of the first challenges for the Consortium is the deep and growing mistrust in American culture regarding medicine, health, and personal data.

ASYMMETRIES OF TRUST

"Given the political climate," Steve wondered aloud to us, "how do we get people that we want to work with, to not only trust us, but to have trust in data?" In a "post-truth" era of unmoored credulity, of "deep fakes," of purloined and weaponized data, and of ideologically motivated data suppression, these factors coalesce to undermine public confidence in how institutions manage their data. In short, there is a lot to be dubious about when it comes to data. Building public confidence in data—that data are accurate, transparent, privacy protected, and are shared with the knowledge and agreement of data subjects—is an ongoing challenge. At the core of Steve's question is a nontechnical but technologically mediated relationship around trust. These questions on how to cultivate trust in data, to encourage people to see data as an objective instrument that can spur a desired change or outcome, form part of wider conversations about trust in institutions, in medicine, in technology, and in science. And these conversations about how to cultivate trust have been around for centuries, if not for millennia. From the Hippocratic oath to the Belmont Report (1979), which protects human subjects in medical research, medical ethical codes are the result of ongoing debates about how a doctor must treat her patient.[15] Because it is trust, or rather, the traumatizing and deadly betrayals of trust, that led to the codification of medical ethics governing all aspects of clinical care and research.[16]

Once the word "trust" is out there on the table, our discussion about data extends to considering the role of universities. Steve pointedly reminds us that institutions such as CMU have their own histories of research duplicity and betrayals of public trust. Many American universities, and their medical schools, have long and fraught narratives of instilling mistrust among the communities that they claim to serve, in particular Black communities. CMU is based in a majority Black city. Some of the more sordid histories–such as the Tuskegee Institute's role in the US Public Health Services' syphilis study, the University of Pennsylvania dermatologist Albert Kligmen's experiments on imprisoned Black men at Holmesberg prison, and the Johns Hopkins University researcher and oncologist Dr. George Gey's development and commercialization of the first successful human cell line from cancerous cells taken from patient Henrietta Lacks without her knowledge or consent—are well known and have shaped legislation and research ethics for much of the previous century (Caplan 1992; Hornblum 1999; Landecker 2000).[17] Past injustices intervene in the present for many university-neighborhood partnerships, despite community demands for and institutional attempts toward reparative programs to restore trust. As bioethicist and historian of medicine Harriet Washington has argued, "Tuskegee" is overburdened as the major cause of mistrust of medicine in the United States among Black Americans (Washington 2021). Rather, it is the combination of knowledge of this history with the ongoing systemic racism and bias many Black Americans experience in their medical encounters—from Black patients being less likely to have their doctor recognize their level of pain, to devices, such as the pulse oximeter, that don't function as well on darker skin—that is at the root of mistrust (Feiner, Severinghaus, and Bickler 2007; Hoffman et al., 2016). While researchers and practitioners in public health, medicine, and healthcare have been sounding alarm bells for more than one hundred years on the dangerous health inequities that Black Americans and other communities of color face, it was only in the wake of the COVID-19 pandemic and the global protests against anti-Black violence and racism, that the American Public Health Association, among others, declared racism a public health crisis (Devakumar et al. 2020).[18] Deep-seated, systemic racism persists throughout the American healthcare and medical systems, and despite ongoing efforts to address endemic biases, only seismic cultural shifts across all sectors of American society will change that fact (Roberts 2011; Washington 2006).

In medicine, many of the relationships that patients form with their doctors, nurses, radiologists, phlebotomists, and others—all of whom are allowed direct access to a patient's body in order to provide medical care—are face-to-face, trust relationships. These trust relationships in medicine pivot on an asymmetrical expectation of intimacy, which is embedded in the patient-clinician relationship a priori. A patient's previous experiences with healthcare, as well as their socioeconomic status, race, and gender identities, and sexuality, and other factors such as ideological or religious beliefs, together influence their willingness or reluctance to trust the doctor or the larger healthcare institution (Hall et al. 2002).

Patients encounter strangers, or near strangers, throughout the clinic. From the moment they enter the door, they are aware of this skewed relationship in every encounter. Patients are expected to trust the stranger who walks into their hospital room with a needle and an IV bag full of what might be saline solution, and asks them to roll up their sleeve or open their gown. Patients are expected to comply. This system preemptively demands a patient's trust on all counts: trust that the medical assistant knows that the needle is sterile, that the IV bag contains the correctly prescribed fluid, and that the assistant has double-checked the patient's chart to make sure that this is the correct room and the correct patient. The patient must trust not only the stranger standing before them with the sharp needle, but also the systems and infrastructures of the entire medical institution. They must trust that all of it is safely "working" and has the patient's care as the central concern (Meyer et al. 2008). This trust must be given without being earned. While a patient might have great confidence in her doctor, that trust may not extend to the larger healthcare system (Ward 2017). Knowing whom and when to trust can be a huge burden for patients to take on, often at the most vulnerable times of their lives.

Patients who mistrust their doctor or the overall healthcare system tend to have poorer health outcomes because they are less adherent to medical advice or less likely to seek out health services in the first place

(Hammond 2010). Over the last fifty years, mistrust of health and medical institutions has significantly risen among Americans across most socioeconomic groups. This drop in confidence tracks with larger downward trends for public trust in institutions more broadly across the American cultural landscape (Gallup n.d.). In 1966, polling data of US residents revealed that more than 76 percent had trust in medical leadership, but by 2017, only 18 percent of surveyed residents had strong confidence in the US healthcare system (Blendon, Benson, and Hero 2014; Khullar 2019). The mistrust of medical and scientific expertise, trending in American society for decades, has only deepened during the coronavirus pandemic. The antivaccination movement, which has been increasingly influential in the last decade or so, was spurred on by conspiracy theories and misinformation planted into the public discourse by the Trump Administration's press briefings. These briefings circulated in a feedback loop between government and these fringe groups.[19]

Trust or distrust in data, especially the sharing of health or medical data, is contextual and relational. While trust in all realms of social life is rooted in relationships, an individual's broader social knowledge and prior expectations influence whether they have trust in another person or institution (Stepanikova et al. 2009). Data policy scholar Barbara Prainsack noted in an interview that when it comes to sharing one's health data with a healthcare provider, trust fluctuates and is a "living, breathing relationship, where both sides do things to earn, maintain, or lose trust of the other."[20]

Because data sharing in healthcare tends to be mediated and indirect—much of our personal health data is produced without our awareness—there is a particularly alienating effect to the data trust relationship. Data about one's health is an abstraction, in the sense that once a patient's body mass index (BMI), for example, is measured and entered into a health record, it becomes a disembodied number, a data point. The patient may no longer have control over the information once it is entered into the electronic health record and may have no say in how the medical community circulates or reuses their collected data. Dhruv Khullar, a physician and healthcare researcher, identified this as one of the valences of data mistrust among patients.[21] Another pressure point of mistrust is the "one-way mirror effect," where patients have little trust in organizations that seem to know more about them than they know themselves as a result of surreptitious data collection and sharing (Ipsos MORI 2017). Wearable devices, data sharing, and predictive models that healthcare settings deploy extend the medical surveillance gaze through data. Many patients already live with the everyday reality of data surveillance in other aspects of their lives, and these devices and platforms, deployed to ensure that patients comply with their medications or medical directives, can further sow mistrust (Auxier et al. 2019; Vogenberg 2009).

The COVID-19 pandemic underlined just how deeply skepticism runs through the American public when it comes to digital surveillance and either governmental or commercial access to personal data. For example, contact tracers working in pandemic hot spots such as Texas and California reported that about half of the people who tested positive for the coronavirus refused to share information with researchers.[22] During an interview, Elya Franciscus, an official with the Harris County Public Health department in Houston, attributed the lack of cooperation in part to a mistrust of surveillance in general, in part to a mistrust of sharing sensitive data specifically with officials, and in part to widespread and politicized misinformation about the pandemic.

> There's so much misinformation being put out right now. So our contact tracers are being—they're being called names. They're being cursed at. Derogatory language is being used because there's been these seeds of mistrust thrown into the community. So when we call, nothing we say can establish that trust where they'll be willing to share information with us.... They think that the numbers are inflated. We've heard multiple people say that we're getting paid to make up results. So it's so difficult to combat all of this information, this mistrust that's being put out there.[23]

Given where the Consortium members (and one observer) sit during this morning meeting, at one of the most powerful and influential medical schools and healthcare systems in the Midwest, we are deeply implicated in this history of mistrust. CM Healthcare is one of the core data partners in the Consortium, and it provides much of the data from the electronic health records of millions of patients who have sought healthcare at one of CM Healthcare's fifteen hospitals. The very data that the hospital system holds are haunted by health inequity, alienation, and racism. Most of these hospitals are located in more affluent and homogenously white neighborhoods, and most of their patients are insured. Once "donated" to the Consortium, these data are intended to do "good work" for communities, especially those communities that are "data poor," such as the neighborhoods that CM Healthcare does not directly serve. To build a data infrastructure that "equitably amplifies community voice and priorities" and to use data to address racial inequity are the Consortium's top mission-driven priorities, according to a promotional presentation Steve developed for a meeting with stakeholders.[24] Toward that end, the Consortium hopes to use data, and the infrastructure supporting it, to build trusting relationships between institutions and the communities that they serve. The challenges of building trust through data, though, are enormous, because mistrust is at the core of the legislative and technical systems of personal data management.

PRACTICING DATA MISTRUST

To understand the everyday experiences of data that divide, betray, and harm, I meet with two public health researchers, Amelia and Valerie. These researchers are associated with Theresa's population health informatics lab, and they talk about how they trust or mistrust health data. Both relay examples from their own experiences as patients rather than as researchers.

On her first visit to a new healthcare provider, Amelia recalled that the nurse taking her vitals and going over her electronic health record remarked, "Oh, I see that you had an abortion ten years ago, is that right?" Amelia did have an abortion about a decade earlier, when she was twenty-nine years old, but she did not see why the nurse asked her such an intrusive, not entirely relevant, and potentially stigmatizing question. She did not have the procedure in the hospital where her doctor was practicing. Due to a variety of reasons, from the laws limiting abortions in the state that she lived in at the time, to her doctor's affiliation with a Catholic hospital, Amelia was forced to have the procedure at a Planned Parenthood clinic about twenty miles away in a neighboring state. Amelia speculates that the Planned Parenthood facility would have developed an electronic health record for her, but as far as she knew or remembers, Planned

Parenthood would not have shared these records beyond the clinic, and certainly not over state lines, especially given the recent and ongoing encroachments from state legislatures and the US Supreme Court on *Roe v. Wade*.[25]

Amelia, who works in the health sciences sector, recalled that the nurse's question felt like a jolt. She told me, "How can I trust my provider to handle my EHR [electronic health record], how did that information get into my new record?" How did her new provider have this information, and why was this particular medical fact about Amelia—that she had had an abortion—the key bit of information that the new nurse wanted to confirm at her very first visit with a new healthcare practice? Amelia described it as a matter of trust: Why did the nurse ask her about that part of her record, and not question or confirm what was more current and relevant? Amelia figured, given how old the information about the procedure was, that the nurse had to have drilled deep into her record to find this piece of her medical history. Why would the nurse be "snooping" around her record like that? How did Amelia's medical history collapse to one procedure done more than a decade ago, so that singular fact about her medical experience would follow her in her record no matter where she received healthcare?

Valerie, a research psychologist who uses data as evidence in her own work, recalled a similar feeling of deep mistrust of health data when she too felt that her record had "flattened" her. That is, her health record didn't reflect her as an individual patient, but rather replicated the social and racial biases of the healthcare system. Valerie recalled an encounter when seeing a new doctor while her regular doctor was on leave. Valerie was about eight months pregnant at the time, and she knew that she was Rh-negative. Her mother had told Valerie about how when she was a baby, she was given a blood transfusion right after her birth. Because of this knowledge about herself, Valerie was always careful to tell her doctor since she would need specialized care prenatally and at delivery to prevent any harm to her own baby. Yet when Valerie went to her doctor's office for a prenatal checkup a few weeks before her due date, she was shocked to learn from the new doctor that her electronic health record contained no information about her Rh-negative status, nor any note or alert for her doctor to provide her with a Rh-factor injection before or during childbirth.

> My [regular] doctor missed that I was Rh negative.... [I learned this when] another doctor saw me and he kept looking and kept looking [at my EHR], and he was like, "You haven't had your Rh factor shot." And I'm like, "What are you talking about?" He said, "You're Rh negative." I said, "Yeah, I told my doctor." And he's just looking at me and said, "I'm going to give it to you anyway."
>
> And when [my regular doctor] came back... I told her, "I told you I was Rh negative." She looked at me and she said, "Yeah. But most Black people aren't Rh negative."
>
> Look, I am not stupid. I knew my blood type.... I didn't know at that time that they had a new treatment. I thought my children were going to be born and they were going to get what I got—a total blood transfusion. That's not necessary anymore, but [my doctor] missed it because most Black people aren't Rh negative.
>
> I'd hate to go back to the Martin Luther King speech and pull out the passages about unfulfilled promises. This country is full of them, and our research community does not seem to be aware of that. And everybody wants to go back to [Tuskegee] for why people mistrust the system. I tell people... that's not what's driving it. [It is their] everyday experiences in these systems where people are not transparent, sometimes they're not telling the truth period. They're covering for each other. And they make mistakes for reasons that just reek of bias.

For both Amelia and Valerie, data contained in their health records that were biased, missing, wrong, dangerously racialized, or irrelevant to the present circumstances breached their trust relationships with healthcare. At worst, healthcare professionals used data to shame or traumatize. In Amelia's case, as well, the question remains open as to exactly how information about a medical procedure done a decade earlier appeared in the just-created electronic health record with a new provider.

This is where the mistrust in data starts, with an individual experience of betrayal or exposure to danger. People learn distrust as much as they learn trust, and cultivating distrust can be an important survival and protective strategy. Distrust can be a check on power, or even a demand for justice (Johnson and Melnikov 2009, 16). The daunting task for healthcare and medicine, then, is to make these institutions worthy of patients' trust: if the institution is trustworthy, then patient trust will follow. For cross-sector community organizations such as the Consortium, or institutions like CM Healthcare that hold personal and sensitive patient data, the challenge around this distrust of data is to prove that they are "honest brokers" or trustworthy stewards of people's data.[26] This aim is more difficult to achieve within a context where the Big Tech giants, like Google and Facebook, have already eroded public trust in how data are collected, used, and shared with no accountability or consent. It is the larger context of surveillance capitalism and data commodification, however, that underlines how uphill the challenge is for the Consortium to cultivate trust through data.

SHARING DISTRUST

How is fundamental data mistrust embedded into the laws, platforms, and practices concerning our data, making some of us worthy or not worthy of trust, or of privacy? Out of a fundamental distrust, legislators drafted the laws that regulate data governance in healthcare and the social support organizations that handle health-related data. The Health Insurance Portability and Accountability Act (HIPAA) was a policy response to widespread insurance and medical billing fraud that were, at the time, largely blamed for driving up overall costs within the healthcare system. When Congress originally passed HIPAA in 1996, policymakers embedded an algorithmic and data-driven logic to detect financial misdeeds in healthcare into the law, making patient data privacy a secondary concern. Neither the title of the act nor in the text of the original law contain a word about privacy. When it was enacted, HIPAA legislation contained no provisions that guaranteed patient data privacy, nor did the law provide patients the right to give informed consent or any control as to how their data could be used and disclosed.[27] At the time, early electronic health record platform technologies were just beginning to make an impact. Many in healthcare could see that the future would be digitized, and they worried about how this would encroach on patient privacy. These concerns spurred healthcare professionals and regulators to develop privacy standards for the collection and transmission of digital records, ahead of the widespread adoption of digital records (Ebeling 2016; Nass, Levit, and Gostin 2009). The Privacy Rule (2003), the Security Rule (2005), and the Omnibus Final Rule (2013)—all amendments to HIPAA—codified these data privacy considerations. But legislators wrote both HIPAA and the HITECH Act, the 2009 legislation that also regulates patient data and digital records, to enable the *disclosure* and *circulation* of private health information via these privacy and data security regulations. Now part of the larger fragmentary data governance framework that contains all kinds of loopholes and gaps, these laws neither protect patient privacy nor prevent harm.

Privacy scholars continue to debate whether privacy can be considered a stable category, or whether it is contextual and fluid, with boundaries changing depending on where and how the data are produced and shared (Ard 2015; Nissenbaum 2010). A general consensus exists that privacy is concerned with an individual's *right to control* how personal information is collected, used, and disclosed, and under what conditions this happens (Allen 2011; Ebeling 2016). Stacey Tovino, a legal scholar who studies privacy compliance in healthcare, centers the patient's perspective when it comes to privacy, which she defines as "an individual's interest in avoiding the unwanted collection by a third party of health or other information about the individual" (Tovino 2016, 33). Because HIPAA focused on securing patient data to *enable* its disclosure, and to prevent fraud above all else, the law did not provision patients with a right to control whether or how their private health information was collected, used, or disclosed, nor did it provide patients the right to determine the conditions of these processes. Breaching data privacy in the context of HIPAA is not a bug but a function of the legislation.

Data ethicist Luciano Floridi posits that data privacy is akin to personhood and the rights that inhere to the control over one's body. The argument that patients should see the monetary value of their data in the same way that hospitals or the pharmaceutical industry do is flawed, and this argument represents a capitulation to biocapitalism and the imperatives of the market, as Floridi notes. In other words, your information is your identity, and anything done to your information is done to you:

> Looking at the nature of a person as being constituted by that person's information allows one to understand the right to informational privacy as a right to personal immunity from unknown, undesired or unintentional changes in one's own identity as an informational entity, either actively—collecting, storing, reproducing, manipulating etc. one's information amounts now to

> stages in cloning and breeding someone's personal identity—or passively—as breaching one's informational privacy may now consist in forcing someone to acquire unwanted data, thus altering her or his nature as an informational entity without consent. (Floridi 2005, 195)

Rather than patient data privacy in healthcare, it is useful to think of "data confidentiality" governance, in which the handling of data receives a distinct type of consideration. Confidentiality entails the obligation on the part of a healthcare professional to protect against and prevent the unauthorized or inappropriate disclosure of information about an identifiable patient, rather than keep patient information private. It is within HIPAA's network of disclosure—a space of confidentiality, but not necessarily of privacy—that much of patient data reside.

The United States does not have omnibus laws that recognize and protect individual privacy, especially data privacy, across all sectors, like the European Union's General Data Protection Regulation (GDPR) and Japan's Personal Information Protection Act (PIPA) (Chico 2018). Instead, each sector regulates an individual's privacy and, in the case of minors, such privacy is also based on the age of the subject. Certain categories of data are considered particularly sensitive and worthy of more intense protections from disclosure, such as information about one's health, finances, or children. While there is a legislative recognition that certain types of data or groups of people require extra privacy protections—as evidenced in laws such as HIPAA, the Fair Credit Reporting Act, the Gramm-Leach-Bliley Act, and the Children's Online Privacy Protection Act (COPPA)—in practice, most data, even the most secured, like private health information, have a way of being exposed and shared beyond the control of the individual or the data steward. Despite legislative efforts to stem the flow of leaked or misused sensitive data, most laws have failed to keep up with the innovations and the monopolistic practices of Big Tech, in large part because lobbyists and lawyers for the information tech industries tend to write or heavily influence draft legislation.[28]

In a survey that asked about digital privacy, a majority of respondents felt that the risks posed by companies' and the government's ubiquitous data collection outweighed any direct benefit to their day-to-day lives. Most Americans do not have any confidence that companies that collect personal data will be good stewards of their information, or that the firms will admit to mistakes when something goes wrong (Auxier et al. 2019). And with every new revelation of Big Tech's data privacy malfeasance, the collective industry tends to respond by doubling down on its power over our data by investing even more in its political influence apparatus. In 2018 alone, Amazon, Facebook, Apple, and Google spent a combined fifty-five million dollars on lobbying lawmakers in Washington, which put them in the same influence-peddling league as the banking and defense industries.[29]

In an effort to stem the flow of private data into the hands of corporations and to rebalance consumers' control over their data, some states have introduced or enacted data privacy laws. These new laws, such as the California Consumer Privacy Act (CCPA), loosely model themselves on the European Union's GDPR, but "loosely" is the operative word. Residents of states with data privacy laws must opt out of data sharing, unlike EU citizens and residents, who are automatically opted-in to GDPR privacy protections.[30] But since these laws in the United States are state laws, residents of other states without such laws have no protection, even if their data reside in data oceans located in a state that does have these data privacy laws, such as California.

Within this context of widespread datafication and marketing surveillance of virtually every aspect of our lives, what counts as "private" has undergone a radical transformation. So has the relationship between privacy and consent, since both the everyday practices of data collection and sharing, and the law, both regularly breach this relationship (Ebeling 2016; Mai 2016; Nissenbaum 2010). For all of these reasons, Steve, the Consortium's co-organizer, is concerned: How can the organization build trust in such a fraught and fragmented data privacy and governance context, especially in the public sector?

BARRIERS TO SHARING

Across the healthcare sector, from hospitals and clinics to pharmacies and insurers, there is broad recognition that patient health data are very valuable, not only in the commercialization of the data themselves, but also in the analyses and models that use patient data to help reduce fines and other penalties that providers can face. Hospitals in particular recognize that the data that their patients produce in the form of electronic health records and claims are prized: for operations, for research, and for financial profit. Some of the barriers to unlock that value in data are technical—issues that have to do with interoperability, or computer code that makes it easier to share data between databases and platforms. The obstacles to sharing may also be structural, as when a department that manages a dataset is reluctant to share that data with another part of the organization.

Privacy

Two of the biggest hurdles to overcome and unlock the value of patient data are patient privacy and consent. For some covered entities—hospitals, insurers, pharmacies, and others whom HIPAA regulations "cover"—the profit potentials of selling patient-derived data for marketing and other purposes can supersede the drive to be HIPAA compliant (Tovino 2016, 44).

To solve the problem of consent, regulators have revised the informed consent processes in medical and human subjects research in the United States. After amendments to the original 1996 HIPAA law, regulations focus on securing data to *enable* data's disclosure to other institutions within the sector, such as pharmacies or payers—health insurers and Medicaid/Medicare—all of which the statute categorizes as covered entities. HIPAA never provisioned patients with a right to control whether or how their private health information was collected, used, or disclosed, nor did it provide patients the right to determine the conditions of how these processes occurred. In the context of data privacy in healthcare it might be more useful to consider that much of HIPAA regulations are concerned with covered entities acting as the stewards, or *honest brokers* as they are often called, of patient data. Patient data, as long as they are shared within HIPAA's network of disclosure, are identifiable and linked back to an individual patient. However, if patient data are shared outside of this safety net, then a process of de-identification (anonymizing) must take place. Privacy regulations identify eighteen points of personally identifiable information that they require to be stripped from a patient's record before it can be shared with a third party outside of HIPAA's network of disclosure. De-identification is one bridge that covered entities cross over to tap the value of patient data. The second bridge is patient consent. First let's look at anonymizing data as a form of trust.

Making Data Trustworthy

> The data are so safe, I can email them to my mother.
> —Theresa describing the privacy and security of synthetic data

During the Consortium's conversation about trusting data, Theresa explains that her lab has been working with a new vendor, SynthDoc, a platform that generates "synthetic data" from the millions of patient health records that the CM Healthcare system holds in its data ocean. Synthetic data are statistically identical to the original dataset but thoroughly "cleaned" of any trace of an individual patient's identity. The deal between SynthDoc and Theresa's lab took a year to iron out, and the partnership is a chance for the overseas startup to break into the US market. SynthDoc will work with Theresa's lab as a test-bed for SynthDoc to refine the platform to be HIPAA-compliant, as well as to make the technology fit the unique value-based care incentives of the US healthcare system. For Theresa's lab, the partnership is the opportunity to unlock and leverage—commercialize—the value of patient data. The synthetic data collaboration will enable Theresa to de-identify patient health record data so effectively that it can never be de-anonymized (reattached to an individual). The synthetic data will exceed HIPAA compliance standards and be "trustworthy" enough to be made into sharable data assets, and thus be monetized into a guaranteed future revenue stream. SynthDoc's marketing emphasizes this asset of the platform, in their words that the SynthDoc platform enables healthcare systems to leverage and "democratize" their patients' data, to empower sharing it across the healthcare ecosystem.[31]

A team of SynthDoc programmers and technicians moved into a suite of offices in Theresa's building, and they regularly meet with lab members to support platform training and research projects. As when medical device companies have sales reps embedded in the hospital, and at times, in surgery rooms guiding surgeons on how to use their company's device,

SynthDoc agents act as facilitators between lab members and the platform (O'Connor, Pollner, and Fugh-Berman 2016). As with the larger data and informatics sector, there is a strong commercialization drive in academic healthcare analytics, but at times it can work at cross-purposes with the drive to produce groundbreaking data research that will improve patients' health. Since Theresa's lab straddles both research and applied health data analytics, some tensions exist between academic health informatics and an outside, third-party vendor such as SynthDoc. While the entire lab has undergone HIPAA training, and all of the active research projects have institutional review board (IRB) approval, all lab members had to sign additional confidentiality agreements with SynthDoc and the CM Healthcare system. For Theresa and the rest of her lab's team of postdocs and students, the SynthDoc coding and algorithms are off-limits; SynthDoc representatives, who become collaborators on all data research projects that the lab develops, mediate every technical detail.

To a population health data scientist like Theresa, trust can look different than it might look to a patient or to the programmers at SynthDoc. For a health informatician, the question may be: How can one trust that the data used in the model are "clean," or accurate, reliable, and a close approximation to the object that they are modeling? Or in the case of synthetic data, how can one trust that the data are completely anonymized, that it is impossible for the data to be reconnected to a patient, and thus are absolutely safe?

Scholars working at the intersections of sociology, software, and data studies highlight the ongoing trust challenges to teams that build artificial intelligence, machine learning, and software systems, especially those that are based on patients' personal data (Golbeck 2009; Knowles and Richards 2021; Pink, Lanzeni, and Horst 2018; Sänger et al. 2014; Thornton, Knowles, and Blair 2021). How does a team "build" trust into the design of the platform from the very beginning? Can trust be scaled throughout the platform? How does the trustworthiness of a platform translate to the end users, and ultimately to the subjects—the public—of these systems? Fundamental to trustworthy software design are transparent communication among collaborators; the ability of team members to track the provenance of each design element—in the case of SynthDoc, the algorithms that synthesize the data; and security in the balance between the openness and privacy of the work being done (Thornton, Knowles, and Blair 2021, 64). While Theresa's lab and SynthDoc work together to build trustworthy synthetic data to "unlock" the research potential as well as the financial value of patient data secured under privacy regulations, the trust between the two groups is something that they constantly negotiate.

During a weekly lab meeting, Jonathan, one of the lab's PhD students in computer science, shared his concerns about trusting SynthDoc. "How do we know that the platform is actually doing what they say it does?" he asked, and added, "I mean, how do we know that the data are safely de-identified?" Jonathan expressed a concern that the group shares: How does the lab know that the platform is producing synthetic data that are completely safe from any possibility of a patient's data being de-anonymized? How does the lab know that the vendor's tool actually does what SynthDoc says it does? And ultimately, how can the lab trust synthetic data? He was so concerned that he told Theresa and the rest of the lab that he had emailed the SynthDoc reps and had yet to receive a response.

Because the platform's intellectual property is proprietary, SynthDoc is unwilling to share its algorithm or method of synthesis, even with its partners. Theresa mentioned that she too had emailed the startup the same question, without a reply. In response to SynthDoc's silence, Theresa asked her postdoctoral researcher, Junling, a computational mathematician with expertise in numerical simulations of particle physics, to write up the theory that she thinks the platform is using to synthesize the data. Junling wrote the theory that, in her best estimation, was the most likely candidate and emailed it to the reps. A few days later, Junling received an email back with some light corrections to her theory, but no confirmation as to whether she had guessed correctly. Theresa and Junling understood they were getting close to how the algorithm actually works, but without a confirmation, they had to trust that the platform's technology was "getting it right" on protecting the patient data privacy.

For synthetic data to become trustworthy, both technically and ethically, it must be a statistically accurate reflection of the original data, and it must also be safely de-identified without jeopardizing patient privacy. Once these criteria are met, Theresa and her lab will be closer to leveraging the value of health data assets that CM Healthcare and the lab own. The next hurdle to overcome is consent for the reuse of patient data.

Consent

Under HIPAA and its Omnibus Rule, patients do not give informed consent to how their data are used, shared, or disclosed. Rather, when a patient enters a medical setting, they are given a Notice of Privacy Practices (the NPP, as it is known colloquially in healthcare settings), a document the Office of Civil Rights within HHS regulates.[32] Often, both patients and practitioners believe that the NPP is a consent form allowing the sharing of data, similar to the informed consent form that a patient will sign before receiving a treatment or procedure. But the NPP is not a consent document; rather, it is simply a notice to patients that the healthcare provider will share and reuse their data in a variety of ways, most of which are not enumerated. The document does not request that patients grant permission to the clinic to share or reuse their data. When patients sign the NPP, their signature is not an endorsement that they have read or understood the NPP, but simply that the health practice gave them the notice. A patient can refuse to sign the document, of course, but then this act will be recorded and kept on file in the patient's electronic health record.

HIPAA was designed to create a regulated information system that would securely disclose patient data for the necessary functioning of profit-driven medicine and healthcare. Essentially HIPAA regulates the security and accuracy of patient data as it passes through the billing and payment infrastructure (Ebeling 2016). A difference exists between the informed consent obtained for medical procedures and research under HIPAA, and the revised Common Rule. The second is a set of ethical standards that fifteen federal departments initially codified in 1991, and it regulates medical and behavioral research (Department of Health and Human Services 2009). Patients do give consent for the reuse of their data once they become research participants, and their health data fall under different ethical protocols. All federally funded human subjects research, and virtually all human subjects research conducted in universities, regardless of funding, abides by the Common Rule. Informed consent obtained from all research participants before participating in or giving samples for research, including any collected data, is fundamental to the Common Rule. But in 2018, in an effort to keep pace with technological changes and research methods, Congressional legislators made revisions to the Common Rule's consent procedures. Research participants now give broad consent—a one-time consent during a project's initial enrollment for use of their tissues or data for the project they are enrolled in, as well as for unknown subsequent studies, without the researchers having to inform the subject.

At the point of initial consent, it is hard to anticipate all the future uses of one's health data. And as data has a way of leaking, migrating, and moving, only to come back to haunt or harm, providing informed consent for one's data can be a particularly fraught exercise. In 2007 the National Institutes of Health (NIH) developed a biobanking initiative, the eMERGE Network, which targets underrepresented populations in genomic research. To understand the potential barriers to participation in this network, researchers asked thirteen thousand clinical patients about their levels of comfort with the study's collection of tissue samples and electronic health records data. Survey respondents noted that one of their greatest reservations about sharing their health records data was losing control over it (Sanderson et al. 2017). Two recent studies, one a meta-analysis and the other a large experimental survey, investigated how patients who were being solicited to enroll in biobanking initiatives felt about giving broad consent—a one-time agreement for their biosamples and clinical data to be widely shared for research purposes or used in secondary research. Most respondents reported their biggest concerns were with the identifiability of their data, the security of their personal information from leaks or privacy breaches, and whether their data might be shared with the government or with commercial entities (Garrison et al. 2016, 669; Sanderson et al. 2017, 415). Those surveyed who identified as members of a historically marginalized group, including Black, Native American, and Latinx patients, voiced these apprehensions more strongly than white patients.

The promises of better health outcomes or an improved quality of life through the mass collection and sharing of biodata can often hide the sacrifices that patients make, both individually and collectively, of their data (TallBear 2013). Starting in 1965, Akimel O'odham ("Pima") residents of the Gila River Indian Community reservation in Arizona were enrolled in an NIH longitudinal study of diabetes and kidney disease. At the time, before the Common Rule was implemented, white researchers argued that the Akimel O'odham, who were experiencing higher than average incidence of diabetes throughout the community, were genetically prone to this condition because they were genetically homogenous. This argument conveniently ignored the obvious social determinants and systemic inequities that were the more likely causes driving up rates of diabetes. Hundreds of years of colonialism, the forced isolation of the reservation, and the strategic undermining of indigenous farming by cutting off water supplies to the reservation limited residents' access to healthy foods and created periods of severe food insecurity. The data collected from the bodies of diabetic residents (and those of their relatives) over the years became the "Pima Indians Data Set" (Radin 2017, 55). A repository of the community's invisible biolabor and their experiences of the slow violence of the reservation, the dataset is used all over the world for tasks well beyond the health and well-being of the Akimel O'odham people (Nixon 2013). The database is now used primarily as a training data set for machine learning in applications as far from community health as any could be, including predicting electrical fires in underground power grids (Radin 2017, 43).

BENDING DATA TOWARD JUSTICE

Open data initiatives and massive biomedical research studies, which produce enormous amounts of personal data connected to individual lives, compound the ongoing and significant issues concerning data sharing and ownership, interoperability and patient privacy in medical research. These endeavors, which involve the sharing of sensitive, personal information about vulnerable populations, pivot around privacy, consent, and the purposes for which the data are shared without either privacy protections or the individual's consent. A workaround for the privacy and consent "clogs" that slow down data sharing and prevent "frictionless" data is for research teams, like Theresa and SynthDoc, to make assets out of both privacy and consent, and thereby make data "trustworthy" and sharable. Yet, the experiences of the Akimel O'odham people, whose biodata and labor were extracted and reused without consent and without their data coming back to them for their direct benefit, is a story repeated over and again, historically and in our current data-centric society.

The Consortium members, including Theresa's lab, aim to overcome these inequities. The revisions to consent under the Common Rule, combined with data engineered to protect privacy, make it possible for biodata to be collected and used in an environment that promotes data stewardship and trustworthiness, and ultimately to produce value: financial for some, justice for others.[33] One way to build trust in data is through including community voices in the data themselves, not just turning to the community as a resource or a source of data collection. Toward that end, the Consortium created a data fellows program and enlisted qualitative researchers of color, who conducted a series of listening sessions with community groups in the neighborhoods CM Healthcare serves. Two of the fellows, Phyllis and Rajaa, used the listening sessions to develop guidelines for the Consortium to embed trust in data from the beginning. To bridge the data divisions, the fellows identified data practices that center community voices and priorities, including data collection that is informed by the ongoing traumas communities experience, and that sees the communities as mutual and equal collaborators. Black data activists' groups, like Data for Black Lives, have demanded that data monopolies such as Google and Amazon return data governance back to Black communities, especially during the pandemic. In the spirit of the disability rights movement's rallying cry "nothing about us without us," Data for Black Lives created a COVID-19 open-data hub in an effort to collect and use data related to the pandemic to save the lives of Black patients (Milner 2020). The Consortium guidelines follow a similar path, demanding that data remain centered on the communities that produce them.

No matter the conditions of collection or use, all of our data, including HIPAA-protected health data, already exist in larger oceans of unregulated (or loosely regulated) consumer, financial, court and legal, and other types of data that are scraped from our day-to-day lives. These data oceans express the values and the ethics that went into their construction (Glissant 1997, 193). Ruja Benjamin, a sociologist who focuses on data rights, digital technologies, and race, argued that if there is to be reparative justice in biomedical and genetics research, we should, as data subjects, be "biodefectors" and participate in informed refusal (Benjamin 2016). Yet under HIPAA and HITECH, and now under the broad consent of the Common Rule, it has become virtually impossible to refuse. We cannot withhold our consent in how our biodata may be used or commercialized, because we are not given the opportunity to consent to how our data are reused in the first place. This is how mistrust is baked into all of the regulations and laws for data privacy and healthcare. The relationship of distrust in data derives from the fact that too often communities and individuals are seen as sites of data extraction and accumulation through their dispossession. The members of the Consortium, including Theresa, want to disrupt this exploitation. In repairing this lack of trust, the hope is for data to bridge these historic divides, and through those bridges, bend the arc toward justice.

3 Collecting Life

Walking downtown along New York City's Fifth Avenue, as I approach Thirteenth Street, a sign in one of the plate-glass windows of the Parsons School of Design's art gallery catches my eye: the Museum of Capitalism. Curious, I decide to go inside to find out what such a museum might contain that would differ from the skyscraper temples to capital that canopy the avenues and streets of Manhattan. The Museum of Capitalism, a traveling art exhibition, is premised on a seemingly simple question: What kinds of artifacts would be collected and exhibited after the collapse of capitalism? The videos, objects, prints, photographs, sound recordings, installations, and other displays work together to speculate upon and imagine possibilities.

Once inside, I pause to watch an eight-minute video by Sharon Daniel, an artist who collaborates with incarcerated women in California's penal system to make videos and art installations displayed in museums and galleries. This video, *Undoing Time/Pledge*, is an interview with a formerly incarcerated woman, Beverly Henry, who spent close to forty years—most of her life—in prison. In the video, Beverly appears middle-aged, has a soft and friendly voice, and seems relaxed; a gentle smile spreads across her face as she speaks. During the interview, Beverly sits with an American flag in her lap and the sound of tearing fabric overlays her voice. While Beverly was incarcerated in the Central California Women's Facility, she sewed American flags for sixty-five cents an hour. During the video portrait, Beverly describes the working conditions she endured as those of a "sweatshop, clear and simple." The state of California, in many ways, collected and controlled Beverly's labor, along with that of the other women who were doing time with her. The artwork's title, *Pledge*, stands in contrast to the Pledge of Allegiance that every school-aged child in the United States can repeat by heart: "I pledge allegiance to the flag of the United States... for which it stands, one nation, under God, with liberty and justice for all." For Beverly and women who share similar circumstances, it is the absence of liberty and justice that makes *Pledge* so poignant, especially given that the United States incarcerates more citizens per capita than any other country. This poignancy threads throughout the museum.

Another object invites my attention, this time a sculpture, *Supermajor*, comprised of nine vintage motor oil canisters bearing old paper labels emblazoned with company names like Pennzoil, Sinclair, and Shell. The cans are stacked in three rows and three columns within a wire frame that holds them in place, a framework much like the motor oil display cases in old gas stations. One can in the center of the stack, a Texaco canister, is punctured and spews out oil onto the platform that the stack perches on, forming a small, glistening amber puddle under the cans. Standing close to it, one can smell the putrid odor of petrochemicals that hovers just above the puddle. The platform is slightly sloped toward a small drainage hole that hides a tube in the base, which feeds the captured oil back into the canisters. The sculpture functions as a fountain, eternally spilling oil. The wall description states that the artist who made the piece, Matt Kenyon, grew up in southern Louisiana, not far from the region dubbed "Cancer Alley," a name earned for the high rates of cancer diagnoses and deaths among residents. The fossil fuel industry's history of environmental, health, political, and global catastrophes features prominently in his work.

Turning around, I spot a small case made of whitewashed wood with a glass top, which is suspended and jutting out of the wall. Inside the case is a colorful collection of about thirty ballpoint pens, lined up side by side. Each pen bears the branding logo of a different drug: Viagra, Lexapro,

Flomax, Crestor. The collection cum art piece is by Cedars-Sinai cardiologist Jeffrey Caren, who over six years collected more than twelve hundred branded pens from pharmaceutical sales reps' marketing visits to his office. In exchange for access to him, Dr. Caren demanded that a rep bring him a branded pen to add to his collection, which he displayed in his office in a wall-size transparent Lucite case. In an interview about his collection becoming a work of art exhibited in the Museum of Capitalism, Dr. Caren said that he never set out to make art out of his collection of pens, but that he intended to "have some fun," possibly at the expense of the pharmaceutical industry's efforts to exert their branding power in a doctor's office and insert a marketing message into the diagnostic relationship.[1] In this sense, his method of collecting and the collection itself are acts of playful defiance against medical capitalism.

Collections can come in many forms. Some are physical objects on display, other collections may contain the living and are rarely exposed to public view, and others may be more ephemeral, like an idea or concept. The art pieces in the Museum of Capitalism describe some of the ways that we can understand collections: in the case of *Pledge,* America's carceral state can be understood as a massive collection of the lives and the futures of millions of incarcerated women, men, and children; *Supermajor* collects a commodity and the slow violence entailed in its extraction and production; in Dr. Caren's pens, a collection of an industry's "goodwill," or the ephemeral value of consumer emotions attached to a brand, that is turned back on itself.[2] The branded pens embody both the physical and the unseen, assembled as a collection of phantom commodities and haunted by the social productive forces that went into their making, all in an ironic gesture to undermine the pharmaceutical companies' effort to gain influence (Marx 1976, 1:126, 163–64).

The Museum of Capitalism is premised on a speculative archive or repository that exposes the present's absurdity and cruelties. Unlike many archives that collect and speculate on the past, the Museum of Capitalism projects its visitors across an imaginary distance into a speculative future—a postcapitalist world—a place and time that affords them a hindsight vision onto the museum's collection of the contemporary. If the purpose of an archive is usually to produce new knowledge about the past, or to reassert the conquerors' power over whose voices the historical record hears, here the purpose seems to be to illuminate and make visible the social conditions that haunt the extractive and exploitative nature of capital: accumulation through dispossession (Foucault 1982; Trundle and Kaplonski 2011). Through its speculative future vision on the present—present conditions, present relationships, present calamities produced by the political economies of late capital—the Museum of Capitalism puts the politics of collections in stark relief by exposing the social conditions and the processes of alienation that went into their making (Pell 2015).

The collections that I focus on in this chapter have a lot in common with those in the museum. These collections are not solid and heavy artifacts like pens or cans of motor oil: they are ephemeral, and even immaterial, in the sense that they are not physical entities. Yet they can have very material effects on people's lives. In many ways, these collections more closely resemble the collections of anonymized life, like Beverly's in *Pledge*. The collections that I refer to are the immense oceans of health, financial, and consumer data that every interaction on- or offline in our data-based society produces or captures. I focus in particular on how patients produce specific categories of these data by drawing them from their bodies and the conditions of their lives, regarding their health and physical wellbeing. These data collections of our data lives, or what I called *lively* data in chapter 1, are more than just stores of inert information. These massive databases are crackling with a lifelike force and the potential to perform work for the institution that controls the data collections.

Once a database or an electronic record platform captures these data, the data rarely stay put. These data will be mobilized to perform any number of tasks and uses, to predict future care needs, or to support decision-making or cost-cutting measures. Your health information will be released from your health record to produce value, to predict, to determine. Data about your health, your identity, and your economic or social class, for instance, can be boiled down and linked to a bit of information as innocuous as the zip code that you live in, which is shared between your doctor, your insurance company, and your employer. This cluster of data can determine what kind of healthcare you receive or do not receive. It can also determine what kind of credit or loan you might be eligible for, which can determine whether you can build wealth through owning your own home. Patients' bodies and the conditions of their health and illnesses produce the data in these collections, which hospitals, doctors' offices, pharmacies, and other healthcare facilities, at least initially, hold. Once these data are extracted from a patient, at first for direct clinical care, the healthcare provider holds and maintains them within their database, increasingly in an electronic health record platform. These collections hold immense value to a variety of interests, such as health insurers, marketers, and medical finance companies, within the healthcare system itself. Yet their worth goes well beyond healthcare, as seen in the "Pima Indians Data Set," the diabetes database of the Akimel O'odham, which is now mostly used as teaching datasets for machine learning (see chapter 2). These massive datasets and records also expose the power relationships of capitalist accumulation of resources and value, which are possible only through the dispossession of those with less power to resist that extraction, as the collections in the Museum of Capitalism show.

When artifacts enter into the collections that archives and museums hold, they enter a graveyard, where they are stripped of the socio-contextual meanings and the powers that gave them life. In the critically anticolonial and anticapitalist essay film *Les statues meurent aussi* (Statues also die) by Alain Resnais and Chris Marker (1953), the camera fixates meditatively on the silent, decontextualized statues, masks, and ritual sculptures that the Musée de l'Homme in Paris holds. European colonizers looted these material objects and tore them from the various African cultures and religions that produced them. These looted objects were killed when they were stripped of the social contexts that animated them with life and power. Close-ups linger on the hollow eyes and mute mouths of the masks and sculpted human figures, interspersed with images of African people swimming, working in factories, or protesting. A narrator argues, "When men die, they enter into history. When statues die, they enter into art. This botany of death is what we call culture.... An object dies when the living glance trained upon it disappears. And when we disappear, our objects will be confined to the place where we send black things: to the museum" (Resnais and Marker 1953). These objects, which colonizing powers hold, are drained of their life forces and become dead things.

Similarly, data collections that corporate interests capture, hold, and colonize are not transparent to the producer of that data or the person from whom the data were extracted, be it a consumer, a patient, a citizen, or a debtor. Here I refer to the collections of experiences, emotions, and conditions that are derived from people's lives on a day-to-day basis (Zuboff 2019). In the data-based society, increasingly our "day-to-day" lives *live* in "data." The genesis of a collection of data is always rooted in the material world—data originate in a measurement or characterization of a body, a geographic phenomenon, an event—but the robust data become flat and decontextualized out of necessity in order to be stored digitally in databases. Once these data are stripped of their liveliness, a process of assetization—what Kean Birch and Fabian Muneisa define as the conversion of resources and things, like digitized data, into assets (2020, 4–5)—gives them a new life. In other words, the data are drained of their original life force and purpose, and become dormant in a company's database, only to be reanimated as they are reorganized, queried, analyzed, repackaged, and deployed for new purposes. Those new purposes usually financially benefit the institutions that own and control the data collections. An example of this process is the data patients produce in the healthcare system. For the rest of this chapter, I will focus on the collections of patient data derived from clinical experiences. These data originate from our bodies, from the status of our health, and are then abstracted quantitatively into databases. Once there, our data are extracted, manipulated, shared, and often monetized in a variety of ways. In this manner, once our data enter into these collections, they live their first afterlife.

OCEANS OF DATA

In his book *Collecting Experiments: Making Big Data Biology*, Bruno Strasser, a historian of the life sciences, delineates the nature of the collections that biological sciences developed and used throughout the twentieth century and how these collections of specimens in drawers and cabinets became the massive datasets and databases that feed bioinformatics research in the twenty-first century (Strasser 2019). Some of the collections he studied were microscopic but physical in nature, such as the original 574 live bacteria strains that established the first public museum of living bacteria in what is now the American Museum of Natural History in New York City. Another example are the blood banks and protein crystallographic databanks, which, with the bacteria strains, resulted in the development of the digitized genetic sequencing information contained within databases. Through the histories of the women and men working in these institutions who collected, curated, and created systems for sharing the data produced from experiments in biology and the natural sciences, *Collecting Experiments* helps to make sense of the current turn toward data-driven, computational experimentation. The physical and "virtual" collections "were not mere repositories; they were tools for producing knowledge" (2019, 196).

As with the biological sciences, which use collections of data derived from life itself to develop new insights, healthcare and medicine also collect large amounts of diverse data derived directly from human life. The rise in cheap computational power, data capturing devices, and distributed cloud computing enables collections that quantify human life, or as Ramon Amaro calls such collections, "life as statistics" (Amaro 2020). These collections may include census data or household surveys for public health, or the massive datasets collected on the finer details of day-to-day life, from measures of one's sleep, say, to number of steps taken on a given day (Amaro 2020; Leonelli 2016a). The technical affordances that allow for such massive collections are possible only because they are assembled in the sociopolitical.

Healthcare and medicine capture these data from a variety of sources, including but not limited to a patient's electronic health record—which includes clinical, procedural, laboratory, and other types of private information such as a patient's Social Security and telephone numbers—and biometric data collected from the Internet of Medical Things (IoMT). These are the internet-connected machines used by a hospital or doctor's office. They include not only MRIs, ultrasound machines, and heart monitors, but also implanted devices such as pacemakers and consumer-grade wearable devices like Fitbits, which allow patients to upload data directly from the device to their health record. Healthcare systems' conversion of their patient records to digital platforms also helps in part to drive the exponential growth in health data. It is futile to try to pin down the quantity of health data produced daily, much less annually, as these estimates would be outdated as soon as they were published. In 2013, for example, it was estimated that the volume of digitized health data produced and collected globally surpassed 153 exabytes (an exabyte is a billion gigabytes); it was projected to reach 2,314 exabytes by 2020, growing at a rate of 48 percent annually. This growth outpaced other types of collected data, such as phone data (Ebeling 2016, 15). Collections of data measured in exabytes are often called data lakes or oceans because of their massive size and depth (or multi-dimensionality) and the complexity of the data they hold. Very powerful, distributed computing on a scale hardly imaginable until as recently as a decade or fifteen years ago makes these data lakes or oceans possible. Digital media scholar Lev Manovich noted that these data oceans, or big data, which cheap and distributed computing power dually produces and analyzes, derive their meaning necessarily from their size (Manovich 2012).

Once a patient and a healthcare provider have established trust (or resignation, as we saw in the previous chapter), allowing the clinic to extract data, the entity that collected the information at the initial point of contact holds a patient's clinical and procedural information within database repositories. Within a patient's electronic health record, this information can include a patient's personal health information (PHI), an object of regulation under HIPAA; clinical data, such as blood pressure readings and body mass index measures; lab results from blood and radiology; and sometimes genetic information, if genetic tests were done or if the patient shared their genetic information from a direct-to-consumer genomics company, such as 23andMe. As I discuss in more detail in the following chapter, increasingly, other types of data can be included in an individual record. These types of data can include behavioral and "social determinants of health" data; for example, documentation of how much a patient is socially isolated, or if their income is below the federal poverty line, or if they have been exposed to violence or trauma. Within the hospital separate and distinct databases mirror the clinical databases that contain each patient's electronic health record. These databases contain all of the claims data submitted to health insurers and other payers. Since each point of contact with an individual patient during a clinical visit generates a billing and procedural code, claims data tend to be very fine-grained, because virtually every procedure, every action, every pill or needle has a price that is billed for. Because of this level of detail, much can be inferred from claims data. While hospitals take steps to delink an individual patient electronic health record from the claims record, there are inevitably overlaps between claims and clinical data, which can include procedural and diagnostic codes, the patient's name, contact information, lab work done, and billing information.

For the population health informatics researchers, such as Theresa whom we met in the previous chapter, the diversity and complexities of these collections pose their own problems. Data may be sloshing around, as Theresa describes it, in the healthcare system, but if it is not interoperable, or easily uploaded and "read" by the hospital's current platforms or by a third party's database, the data are virtually useless. For instance, across the fifteen hospitals and clinics that comprise CM Healthcare, the CMU hospital system, there are fifty legacy electronic health records systems. These systems all contain data that cannot be easily integrated into newer electronic health record systems or used for other purposes, such as for the research that Theresa's lab conducts. The institutional efforts required to integrate these data across platforms are enormous and expensive. Because CM Healthcare is a research institution as well, with affiliated genetics, oncology, and other biomedical research centers within the medical school, each department's new data and knowledge can be difficult to share across platforms and with other departments, not to mention outside the CM Healthcare system.

For patients' data to produce value for the healthcare sector as well as Big Tech—in the immediate and the longer term—they must be mobilized for a variety of purposes well beyond patients' direct care. The very data that patients give up in exchange for medical services produce value only after they are released to the health insurer for reimbursement or to informatics researchers who then feed the data into a machine-learning algorithm in the hopes of predicting a health risk or outcome.

These data are trawled out of the data oceans to be reanimated and made to perform unimaginable feats through algorithmic physics, but they are used primarily to produce "efficiencies" —another way of saying cost-cutting—or to produce financial value. For instance, health insurers hold massive amounts of patient data, specifically claims data, that third parties or the employers that subscribe to health plans on behalf of their employees can utilize. Castlight, for example, is an informatics company that helps employers track, visualize, and predict their employees' health behaviors and needs through health insurance claims data. Through data-sharing agreements made with the employer and the health insurer, the Castlight platform produces visual data graphs that can show, for example, how many employees are pregnant, trying to get pregnant, or may become pregnant.[3] With Castlight products and other tracking apps that allow health data sharing between employees and employers, instead of some familiar but illegal practices used to surveil employees' bodies at work, such as asking female employees if they plan on getting pregnant, employers can use insurance claims data to identify potential drains on productivity and future losses. In this way employers can anticipate, and possibly curtail, the added expenses of sick pay or maternity leave coverage.[4] These collections of data produce additional value when they end up in databases outside of healthcare, such as those that Google or Experian own. In fact, data are shared with or enhanced by third-party commercial interests, often considered outside the traditional healthcare space. In the data-based society, however, the distinctions and any regulatory firewalls between commercial sectors are now completely fluid. The primary way that these vast oceans of data produce value is through assetization. Assets are objects that can be:

> owned or controlled, traded, and capitalized on as a revenue stream, often involving the valuation of discounted future earnings in the present—it could be a piece of land, a skill or experience, a sum of money, a bodily function or affective personality, a life-form, a patent or copyright, and so on. . . . Assets can be bought and sold, yes. But the point is to get a durable economic rent from them, not to sell them in the market today; here we use the term rent to mean the extraction of value through the ownership and control of an asset, which usually entails limiting access to it. (Birch and Muniesa 2020, 2)

The data collections the US healthcare system uses are increasingly commercialized or traded in ways that patients, health plan subscribers, and healthcare practitioners have found troubling, and that some are pushing back against.[5] It is through data-sharing agreements that the collections of our data go on to live afterlives as objects of value production for, or assets that return rents to, Big Tech monopolies.

CANARY IN THE DATABASE

> I must speak out about the things that are going on behind the scenes.
>
> —Anonymous whistleblower who exposed Google Health's Project Nightingale

In an anonymously uploaded video, a Google employee whistleblower records their computer screen as they click through several PowerPoint presentations outlining a new partnership—code name Project Nightingale—between Google Health and the United States' second-largest healthcare provider, Ascension Health.[6] This partnership builds AI and machine-learning products to help Ascension better manage its clinical care and financial operations. The slides, dated between May and August 2019, appear to be confidential materials that not only outline the nature of the partnership between the two companies, but also describe the data-sharing structures, platforms, tools, and analytics that Google will build based on the more than fifty million patients from Ascension's nonprofit healthcare system, which sprawls across twenty-one states and the District of Columbia, and is headquartered in St. Louis.[7]

Throughout the silent video, the whistleblower annotates the slides of the various presentations, to highlight for the viewer—presumably an imagined patient who doesn't know a lot about what happens to their medical data—what is most concerning about the deal. On one slide, the whistleblower notes that Ascension has already moved millions of patient records to the Google Cloud platform. The annotation emphasizes that these records still contain all the patients' identifiable information, what under HIPAA are called the patient's protected health information or PHI. Clinicians, patients, and those in the data industry invoke both the acronym HIPAA and, to a lesser extent, the term PHI, as shorthand for patient data privacy and security in healthcare.

Under HIPAA's privacy regulations, before data can be shared with third parties outside the healthcare system, the covered entity—the hospital or healthcare provider that holds patient data—must strip the records of the PHI. On another slide, the whistleblower notes that Google will build four technical abilities or uses, based on Ascension's individually identifiable patient data, including data-mining, machine-learning, and deep-learning algorithms. Since the data are identifiable, any Google employee, or more worrying to the whistleblower, a hacker, can easily search the records and see detailed information about individual patients. These data include their names, ages, home addresses, diagnoses, lab results—essentially all the information about a patient that HIPAA ostensibly protects. Once the data are in Google's possession, the company can then sell or share these data with other third parties outside Project Nightingale, or Google could use the identifiable patient data to target advertising based on the health conditions visible within the patients' records. When a patient initially sought healthcare from one of Ascension's hospitals or clinics, that patient would have signed a piece of paper or a computer screen acknowledging that Ascension might share their health record with a business partner to improve "healthcare operations." That patient, however, would not have given consent for their data to be shared; since 1997, informed consent for the sharing of patient data was eliminated under HIPAA. Now patients can only acknowledge that they have been informed that their data *will* be shared.

Soon after the whistleblower leaked information about Project Nightingale to the *Wall Street Journal*, they published an anonymous commentary in the *Guardian*, one of the news outlets that in 2013 published the groundbreaking leak by former National Security Administration (NSA) contractor Edward Snowden concerning the NSA's secretive and massive surveillance of American citizens. In their commentary, the Google whistleblower explained that after their initial enthusiasm about working on Project Nightingale, their worries about what Google was doing with the identifiable health data of more than fifty million Americans grew. Their colleagues at both Google and Ascension secretly shared these worries. The whistleblower's biggest concern was that the established HIPAA protocols concerning patient data privacy, in the hands of the search giant Google, were not strong enough, nor was Google applying these regulations carefully enough: "Data security is important in any field, but when that data relates to the personal details of an individual's health, it is of the utmost importance as this is the last frontier of data privacy."[8]

After the news story reported the whistleblower's leak about the Project Nightingale deal, Google issued a public statement concerning its use of the identifiable health records, arguing against reports that the data-sharing agreement violated HIPAA privacy regulations. In the statement, posted to the Google Cloud, Tariq Skahaut, president of Google's Industry Solutions and Products division, explained that Google has a

> Business Associate Agreement (BAA) with Ascension, which governs access to Protected Health Information (PHI) for the purpose of helping providers support patient care. This is standard practice in healthcare, as patient data is frequently managed in electronic systems that nurses and doctors widely use to deliver patient care. To be clear: under this arrangement, Ascension's data cannot be used for any other purpose than for providing these services we're offering under the agreement, and patient data cannot and will not be combined with any Google consumer data.[9]

This statement was an effort to assuage any public concern or mistrust of the company's use of patient data.

Within the strict confines of regulations, yes, Project Nightingale technically complies with HIPAA's privacy rules concerning patients' protected health information. Under the Omnibus Act of 2013 that amended the original HIPAA privacy and data security regulations, business associates of covered entities can use PHI data if this work helps or enhances the healthcare operations of the covered entity. Google would be considered a business associate of Ascension, as the press release stated. Because the deal used patient data-driven AI and machine learning to improve healthcare delivery, Ascension and Google both argued publicly that the deal was within the scope of approved data-sharing uses and complied with HIPAA regulations. In response to the coronavirus pandemic that took hold in the United States in spring 2020, the Office for Civil Rights (OCR) of the HHS, the patient data privacy enforcement agency, suspended aspects of HIPAA that concern business associates' uses and disclosures of patient data, and that concern telemedicine (Office for Civil Rights 2020). The federal government's lifting of privacy regulations pertaining to patient data affirmed Google's defensive position in regard to breaching patient privacy and public trust.

These commercial relationships between nonprofit healthcare organizations and information companies like Google, Amazon, and Experian, which are hidden from public scrutiny behind data-use agreements, nondisclosure agreements, and confidentiality contracts, are fundamental to the data-driven healthcare system in our data-based society. News of these kinds of data-sharing arrangements in healthcare tend to elicit shock and anger when exposed, and although these deals seem novel and recent, they are neither. The trade in patient health information, identifiable or de-identified, clinical or not, between healthcare providers and commercial interests outside of healthcare occurs routinely, frequently, and often secretly. As more healthcare systems become digitized, these commercial contracts increase. For instance, Google and the United Kingdom's public National Health Service (NHS) ran into a backlash from patients when similar data-sharing agreements were revealed between the NHS and Google's London-based AI subsidiary, DeepMind, in 2016. The Information Commission Office, the UK regulatory body that oversees data governance, declared the transfer of 1.6 million identifiable patient records to DeepMind unlawful and subsequently killed the project. Yet three years later, these NHS patient records were transferred as planned, after DeepMind ceded all its work in healthcare to Google Health, based in California. These NHS patient records, now under control of an American company, were developed under the European Union's data privacy regulatory framework and the NHS's patient consent protocols.[10]

DATA IS POWER, BUT DATA WILL NOT EMPOWER US

Before the global coronavirus pandemic made most events virtual, I traveled to Washington, DC, to attend Health Datapalooza. Academy Health, one of the more influential organizations that promotes open science and data-driven healthcare policy in the United States, sponsored this annual meeting. The Datapalooza moniker riffs on the once defunct, and now resurrected, music festival Lollapalooza. But the Datapalooza name gives the Washington, DC event an anachronistic feel of the late 1990s, which is fitting for what is, essentially, an expensive lobbying and influence-peddling event for the health data industry. The year that I went, Joe Biden gave the keynote to the standing-room-only crowd.

At many of these tech industry-focused conventions that I've been to over the years, the bluster about the promise and earth-shattering potential of health data innovations is always for the "good"—for the good of patients, the healthcare industry, and ultimately, society. I heard or read this effusive rhetoric about the promissory good that data can do throughout Health Datapalooza, from the branding on the tote bags, notepads, and conference programs given to attendees, to the posters and presentations of the panelists and keynote speakers. During a panel on patient privacy, for instance, an audience member drew a parallel with fossil fuels by declaring that patient data is the "new oil" driving twenty-first-century healthcare, and that, because of how valuable it is, it should be liberated from its confinement in patients' electronic health records. Patient data is worth so much to medical science, he went on, it should be considered a natural resource; for such a crucially valuable supply of data to remain locked up in databases because of privacy regulations is unconscionable, even unethical. These data, the speaker argued, must be freed to do "good."

Yet this data-based "natural" resource, once it is extracted from patients and undergoes the assetization process, is often delivered right into the hands of data monopolies. Once private health data are mined from health records, they are refined, made into assets, and utilized to produce value for shareholders and investors in capitalist medicine and Big Tech. Far from being liberated to do good for patients' direct benefit by delivering on the promise of better health, patient data within the context of American for-profit healthcare are considered the innovator's asset, closely guarded or sold as a commodity. In a certain way, then, patient data becomes a "good" in the economic sense of an asset that holds potential or future capital value.

The collection of massive amounts of data about us not only has direct impacts on the quality of our lives, but also disempowers us, in direct contradiction to the oft-repeated claim that data is empowerment. How does this happen? The legislative processes—which far from being democratic or speaking for the voiceless, tend to favor the powerful—disempower us. Most of the patient data sharing within the healthcare environment remains hidden from us because this collection goes on largely without our knowledge or consent.

These data-sharing deals are always done in the best interests of the corporations involved and often push at the edges of what is legal and what might be considered ethical. They differ markedly from the kind of data-sharing that the Consortium proposed in the previous chapter. When these behind-the-scenes agreements are exposed, the official public pronouncements are often that these deals will empower patients, make patients healthier, and improve healthcare delivery. These public-private data-sharing deals perform operations in the best interests of the company while avoiding public discussion of the underlying threat—in the case of Project Nightingale, the companies sought to smooth the ruffled feathers of Google users, who by now have ceded to the company's monopolistic control of their data.

These corporate communications—uttered in a bid to regain public trust after secret deals are exposed publicly—inevitably perform a double-think speech act. The maneuver works on two registers. The first is the corporate justification of receiving identifiable patient data because patients already consented, under HIPAA, to third parties using their data. However, patients do not give consent to what happens to their data; patients simply acknowledge that a repurposed use of their data may occur. Second, the corporate speech works as a gaslighting technique. Contrary to patients' feelings of disempowerment when they learn that their identifiable private health data have been traded, sold, and possibly exposed to unknown parties, corporate public communications will inevitably argue that these bargains are in patients' best interests, because ultimately their data will be used to improve clinical care.

Patients are disempowered legislatively through the laws and regulations concerning the collection and subsequent dissemination of electronic health records, patient health and medical data, and online patient portals that give limited access to patient records. Policymakers use the rhetorical device of *empowerment*—a term they borrow directly from the industries that people elect them to regulate. Policymakers' activities, however, in fact disempower patients with regard to access to or control of their data, and certainly do not change patients' relationship to their data. Rather, policymakers legislatively cede control over patient data back to commercial interests. This action ensures that powerful interests operating within the data economy—those companies and corporations that profit from analyzing and trading in patient data—have frictionless access to this core asset.

Consider the Health Information Technology for Economic and Clinical Health (HITECH) Act, passed during the Obama administration.[11] HITECH focuses on patient data and their production, access, innovation, and movement within the American healthcare system. Far from empowering patients through access to and control over their personal health data, HITECH ensures that powerful corporate interests in the consumer health data and information technologies industries can access, control, and assetize patient data, because the industry lobbyists wrote the legislation. In what Sheila Jasanoff described as a "regulatory retreat" of the state that began more than thirty years ago in the United States, both the act and the bill demonstrate that the legislative processes that regulate patient data have been handed over to third-party interests. The aim is not to improve health outcomes for patients through access to health and medical data, but to enhance bottom lines for the ancillary healthcare industries that write the regulations that lawmakers pass (Jasanoff 2011, 631).

Despite the ongoing legislative gridlock over providing a system that offers better healthcare at lower costs, it is the legislative process itself, where the healthcare industry has embedded itself in political decision-making, that is creating the system's out-of-control costs.[12] In a study of hospitals that invested resources into lobbying efforts, the hospitals realized direct benefit from certain legislative provisions, but also showed a 25 percent increase in their spending, without producing improved patient health outcomes (Cooper et al. 2017).

Within American healthcare, corporate control and free market ideology have long been the reality, and regulators such as the Food and Drug Administration (FDA) have historically appointed industry leaders from big pharma and big agriculture to head up the agency. The FDA has the reputation of being a "revolving door" between regulators and the industries they oversee (Bien and Prasad 2016). Some of the high costs of the healthcare system come from the legislative process itself; the healthcare industry, which includes pharmaceutical companies and information and medical technology corporations, spends more on political contributions to Congress than even the defense lobby.[13] In 2016 alone, the healthcare lobby made more than five hundred million dollars in political contributions to Congress. Healthcare lobbyists also write the bills and legislation that Congress makes into laws that regulate the healthcare industry.[14]

THE HITECH ACT AND DATA OCEANS

Out of the smoldering wreckage of the global economic crisis of 2008—a crash instigated by Wall Street's speculative gambling with ephemeral debt-based assets (primarily mortgage-backed derivatives) and built on obfuscation and fraud—a phoenix rose for the American healthcare system. Or at least that's how it's been widely marketed to American patients and healthcare practitioners. As the previous chapter focused on HIPAA and the distrust embedded in the law, here my interest is in how the industry created, controlled, and operationalized massive collections of health data to produce value for those companies, such as Experian or other third parties outside of healthcare, that have a monopolized control over personal data. The industry accomplished this end by leveraging legislative and technical power and passing laws favorable to data companies. The resulting data monopolization dispossesses us not only of our data, but also of the political power to control our data, since it cuts us out of the legislative processes where we could determine what happens to the information that has repercussions on virtually every aspect of our lives.

The HITECH Act, passed in February 2009, was part of the expansive American Recovery and Reinvestment Act (the Recovery Act), a bill intended to stimulate the US economy after the worldwide economic collapse. By favoring certain sectors of the economy understood to encourage growth—such as healthcare information technologies and allied industries—with stimulus spending and advantageous legislation, President Obama hoped to pull the American economy up from its nadir in the worldwide recession without the austerity measures that other countries, such as the United Kingdom and Greece, had chosen (Bivens 2016).

If the "digital revolution" and the promise of data to transform healthcare—or *disrupt* it, to borrow the Silicon Valley buzzword—define the twenty-first century, the passage of the HITECH Act was intended to reform the fractured and expensive American healthcare system. The HITECH Act's proponents widely touted the old, familiar trope that digital health information would become a tool for patient empowerment, and once empowered, patients could take responsibility for their health. Electronic health records would bring us closer to realizing the promises of personalized medicine on a massive scale, which, of course, would ultimately lead to universally better health outcomes for patients.[15] If that were not enough of a promise of the disruptive power of HITECH, all these gains for patients would also result in more efficient healthcare delivery, and ostensibly, a cheaper healthcare system.[16] All this would happen through the transformative powers of digitized health records and data.

The HITECH Act mandated that all healthcare systems—from the smallest family practice to the largest for-profit corporate hospital systems—demonstrate a meaningful use of healthcare IT through a transition to a fully digital health records system. Through the act, Congress also earmarked two billion dollars in federal funds to support hospitals' and practices' transition to electronic health records (Stark 2010, SP26). A series of incentives enforced the mandated demonstration of meaningful use until the 2015 implementation deadline, after which punitive measures followed these incentives. A significant way for healthcare practices, covered entities under HIPAA, to demonstrate meaningful use was through the provision of online portals to patients, where patients have limited access to and can manage, albeit in a perfunctory way, their own digital health record. Several years after HITECH's rollout of its meaningful-use mandate concerning patient portals, in a survey of American Health Information Management Association (AHIMA) member hospitals and clinics, while 87.5 percent of practices had implemented an electronic health records platform, only 38 percent had made a patient portal available online (Murphy-Abdouch 2015, 3). Of those practices that provided online patient access to records, fewer than 5 percent of their patients utilized the portals (2015, 3). We should understand even this limited adoption not as a way of empowering patients to take control over their health and care via access to their data, but rather as a concession to patient advocates demanding more control over their data and privacy.

Neither patients nor even providers, as promoters promised, necessarily felt revolutionary changes and benefits emerge from the HITECH Act. The clear winner was the industry positioned to gain the most financially through the digital medicine revolution: those companies that make the tools for, records platforms for, and data assets out of digitized patient information, such as Epic and Cerner.[17] For many health practices in the highly fragmented, capitalist American medical system, the conversion to electronic health records was a huge and expensive enterprise, even with the financial incentives the bill offered until 2015. Some larger healthcare systems, such as Partners HealthCare, a system of hospitals and clinics spread across several New England states, spent $1.6 billion to upgrade their digital systems to the privately held electronic health records platform Epic to comply with HITECH by the 2015 deadline.[18] The largest chunks of this eye-watering $1.6 billion were lost patient revenue during the rollout phases of the platform and all the tech support personnel hired to implement the platform. Smaller private practices and family doctors, many of whom simply lacked the capacity or capital to transition decades-worth of paper-based patient records to digital platforms in the relatively quick turnaround HITECH required, were forced to retire and shutter, or sold their practices to larger healthcare systems that had the capacity to migrate to digital platforms (Ebeling 2016).

Aside from the immediate financial benefits generated for electronic health record platform companies, HITECH also created massive pools, or oceans, of patient data that hospitals, clinics, public health offices, and health insurers mine for myriad purposes. The notion that data is a natural resource to be tapped originated in the consumer data industry, yet many in medicine and healthcare see patient data as the "new oil." It is understood as a resource, a well of pluripotent information, that clinicians, researchers, administrators, and commercial actors need access to in order to tap the value locked inside (Meister, Deiters, and Becker 2016).[19]

Once these data are privacy protected and compliant under HIPAA and HITECH regulations, they become data assets that covered entities own and can trade and sell. The healthcare industry's view that patient data are in fact company assets that produce financial value intensified with the *Sorrell v. IMS Health Inc.* decision. This 2011 Supreme Court case determined that healthcare facilities, or covered entities, in possession of patient data, *own* those data, and therefore, certain uses and disclosures of those data are considered protected speech under the First Amendment (Boumil et al. 2012; Ebeling 2016).[20] After this decision, these data were seen as productive commodities that, once unleashed from electronic health records and other data platforms, could generate value back to covered entities.

Because HITECH helped to create massive sets of patient data, and *Sorrell v. IMS Health Inc.* enabled new third-party uses of data, these oceans of data became raw resources that companies could tap and transform into data commodities for myriad purposes. These purposes, including the complex insurance billing system, were unrelated to the direct care of patients. Patient data, once considered dormant or dead after a covered entity initially used the data, become reanimated with new life through innovation. These data become the lively data mobilized outside of the health records as highly valuable and coveted assets—medical, political, and financial (Ebeling 2018, 2016). Operationalized in the right way, patient data can be used on many fronts: to prove to Centers for Medicare and Medicaid Services (CMS) regulators that a hospital's readmission rates are going down, and thus avoid huge fines; to show the ability to conduct clinical trials *in silico* (that is, by algorithm); to bill payers; and to sell de-identified data to third parties, such as IQVIA, a health data furnisher, which sell these data for marketing purposes to pharmaceutical companies and others within the allied healthcare industry.[21] In this regard, patients—or *consumers*—are paying twice: once when they pay for their healthcare, either out-of-pocket or through a payer; and a second time, when they are dispossessed of their data, which is sold to third parties, only to return back to patients in more costly drugs, healthcare devices, and services.

ACCESS TO HEALTH RECORDS—FOR PATIENTS OR COMMERCIAL INTERESTS?

In late January 2018, three powerful corporations in the United States—online retailer Amazon and financial services firms Berkshire Hathaway and JP Morgan Chase—publicly announced a partnership to create a healthcare company for their employees. Some observers claimed that this deal could, again, *disrupt* the healthcare market by leveraging the information and data technologies that these companies already own and applying them to healthcare delivery. The partnership, formalized as Haven Healthcare, was short-lived: founded in late January 2018, it shuttered its operations by February 2021. In its early days, Haven garnered much attention both in the national and industry press about the potentials of using the personal data of more than 1.2 million employees to deliver "healthcare efficiencies"—essentially to wring lower prices out of providers.[22] In a National Public Radio interview about Haven, Duke University health economist Kevin Schulman described the possibilities of a company like Amazon providing healthcare services: "So imagine that you had all your data on Amazon—on an Amazon device. When you woke up in the morning, you didn't feel well, you talk to Alexa. And she said, 'You know what, Steve? Maybe you should come and see us or maybe you should get your blood drawn, and I'll set up an appointment for you.'"[23] The interviewer, Steve Inskeep, asked a follow-up question: "Oh, so the smart speaker is going to become your admitting nurse or whatever you want to call it?" Schulman responded, "It absolutely could. And the backbone of this could—would have to be access to really high-quality data that you generally don't have. Your doctor has it or their health system [has it]. One of the things that they're going to have to do is work on *liberating the data* so these services can actually be really impactful for you" (my emphasis).

Schulman's statement reveals a fact well known to corporations working in the larger data economy: the value of patient data, locked up in electronic health records and controlled by HIPAA-mandated covered entities, if unleashed from that regulatory control, promises lucrative returns for commercial interests. Of course, this is always framed as a win-win: data empower patients, at the same time that healthcare delivery costs drop through the efficiencies created by the access to and mobilization of data into the larger data economy. Legislation that regulates health information technologies and patient data necessarily exists within the broader data economy of huge data companies such as Google, as we saw with its Project Nightingale data-sharing deal with Ascension Health, and credit and consumer data brokers such as Experian. These data brokers have core operations ostensibly outside of the healthcare sector, but they nevertheless increasingly position themselves to enter the lucrative market in patient-derived health data (Ebeling 2018; Prainsack 2017; Tempini and Del Savio 2018). Because these companies already sit at the center of the sociopolitical assemblages of the larger data economy, it is only a matter of rebranding or repositioning themselves to be poised to dig into the patient data goldmine. One way these commercial interests reposition themselves closer to the front of patient data is through the outright drafting of new legislation. To demonstrate this point, take the example of US H.R. 4613, the Ensuring Patient Access to Healthcare Records Act of 2017, penned by a very powerful information and data company that for the last several years has repositioned itself more centrally to the health data market.[24]

Despite the stated purpose of the Ensuring Patient Access to Healthcare Records Act of 2017, which is to "promote patient access to information related to their care, including real world outcomes and economic data," most of the legislation's language places the author—a commercial data broker—closer to the center of the network of health data.[25] While parts of the bill do specify how clearinghouses can help other covered entities, such as hospitals, to provide online portals and other data platforms to improve health data access for patients, the bulk of the text expands the clearinghouses' authority. The bill's author achieves this by redefining the commercial data broker as a healthcare clearinghouse and by clarifying the status of healthcare clearinghouses as covered entities, rather than as business associates or third parties. HIPAA defines a healthcare clearinghouse as a public or private entity that processes or facilitates the processing of nonstandard health information data elements into standard data elements that comply under the regulatory regimes of HIPAA and HITECH. For many commercial data companies outside of the healthcare sector, it is crucial to be redefined as healthcare clearinghouses.

The legislation's next goal is to redefine clearinghouses as covered entities. The bill states that the proposed legislation "shall not consider healthcare clearinghouses to be business associates for any other purpose. Such clearinghouses shall be considered covered entities."[26] Under this new regulatory designation, clearinghouses would have the same authority as any other covered entity, including the use and disclosure of patients' protected health information. And as with other covered entities, these uses and disclosures occur without obtaining individual patient authorization. Under HIPAA, patients do not give consent as to how their data is used or disclosed because covered entities are required to give notice and obtain consent from patients only in cases where they use PHI for limited marketing purposes, as discussed earlier. Yet language deeper into the bill reveals that the unstated purpose of the bill is to extend the authority of healthcare clearinghouses even beyond that of covered entities:

> In addition to carrying out claims processing functions, [clearinghouses are to] be permitted to use and disclose protected health information without obtaining individual authorization to the same extent as other covered entities . . . and creating de-identified information as permitted by section 164.502(d) of title 22 45, Code of Federal Regulations (McMorris Rodgers et al. 2017, 8).

Essentially, the bill expands the authority of healthcare clearinghouses to aggregate and share depersonalized information *"without respect to whether the recipient of the information has or had a relationship with the patient"* (2017, 8, my emphasis). Considered in total, the bill gives broad powers to healthcare clearinghouses to access and use patient information beyond the current uses of covered entities, without having to obtain additional authorization from patients to use and disclose data with third parties that may not be directly, or even indirectly, involved in patient care. While elements of the bill aim to provide patients with limited access to their own information, it is mostly in the form of aggregated data, which is unlikely to be of value to individual patients. Patients can request access to their own specific data files, for a fee, but ultimately may not be able to obtain it if the clearinghouse determines it isn't technologically feasible to share the full file with the individual.

The bill, however, gives the data broker full access—because it defines the broker as a covered entity. By sitting at the source of the production of patient information, the broker can maximize its profitability and control how data are made into assets. Far from empowering patients through digital health information technologies, the bill potentially deepens patients' alienation from their data. Again, this bill obscures the original ownership claims. Patients are not the owners of the valuable data that their bodies and health produce; rather, the prospectors of patient data own and control that data.

The notion of patient empowerment through health data—which helped to rally public resources in the early years of digitalization of clinical and medical billing data—is now no more than an empty platitude, a marketing bid to sell the public on multinational corporate interests' widespread collection, ownership, and commodification of their health data. Policymakers who pass legislation that supports the monopolistic ownership of health data—all under the marketing guise that data will deliver patient-centered care or enable patients to take "control" over their health—enable this private accumulation of patient data. The promise that digital technologies, open data, and data analytics will solve some of the most vexing questions in medical science, and will make healthcare more accessible, affordable, and equitable, in fact hides the ongoing structural inequities and injustices in healthcare systems. Moreover, as Marine Al Dahdah shows, technologically based, digitally mediated health interventions, such as the mHealth initiatives deployed in Ghana and India that she investigated, double as opportunities to open new target markets and sites of data extraction. These new markets add to the data monopolies that the world's most powerful healthcare and tech companies, such as Microsoft, IBM, and Google, control (Al Dahdah 2019).

Effusive promotion of meaningful use of digital health data relies on two technologically deterministic themes. The first theme is that the massive collections of digitized patient data will create efficiencies and better health outcomes for patients and will lead to data-driven, personalized medicine. The second theme is that patients can take control over their health (but not necessarily their interactions with the medical system) by having digital access to portions of their records. For all the discussions about how electronic health records are a disruptive technology, these discussions rarely challenge the fundamentally skewed relationships of data ownership, assetization, and commodification, or of consent. It is the data innovators (corporations), and not the data producers (patients), who retain the ownership stakes and realize the value of data assets. In many ways, through both the Ensuring Patient Access to Healthcare Records Act and the bill it grew out of, patients are no longer consumers: they are products, or data assets, that provide a continuing source of value for the digital health data economy.

These collections of our most intimate data—information about our health and what we are made of, our bodies, cells, organs, blood, DNA—are really repositories of our desires, fears, aspirations, hopes, traumas, and dreams. Philosopher of science Sabina Leonelli, in her book *Data-Centric Biology*, advocates for understanding data as a relational object, as something that can be understood only in relation to other things, not as something that is an independent index of reality or the truth of something outside of the context in which it is being considered (Leonelli 2016a). From the latter perspective, Leonelli notes that data analysis "involves uncovering which aspects of reality they document, and their epistemic significance stems from their ability to represent . . . reality irrespectively of the interests and the [social] situations of the people handling them" (Leonelli 2016a, 79). We should consider instead, she urges, that data are not the reflection of some outside reality, but rather bear the marks and significance of how they are collected, what relation they bear to the collector, why they are collected, how they are made into assets that produce value, and how they will be used as objects of analysis. Leonelli therefore defines data in relation to the interpretative contexts for which they are used, rather than by some intrinsic or ontological property they may have.

For the purposes of our understanding this first afterlife of data in healthcare databases, then, these data take on the properties of medical capitalism. This capitalism collects and analyzes healthcare data within a system of procedural costs and billing. Anmei, a nursing informatics researcher and coworker of Theresa's at CMU who conducts surveys with health practitioners on usability issues and adaption to electronic health records platforms like Epic, emphasized that electronic health records were never developed for patient care, but for billing. In an interview, Anmei noted that the electronic health records' purpose is to make billing more efficient and accurate, "so that's why it's not designed for clinicians to take care of patients. That's also why clinicians are not happy, because it's not helping them [to take] care of the patient." We can trace the fact that electronic health records are built for billing and not care directly back to the passage of HIPAA in 1996. The data collected and maintained in the system were not to directly benefit patients, or even to make workflows easier for clinicians, though these could be side effects. The point was to enable more and higher reimbursements for procedures from insurers and payers, and to cut down on insurance fraud. In the surveys that Anmei conducted with clinicians who use a proprietary electronic health record platform at the CM Healthcare hospitals, many respondents perceived that the decision to use this particular platform was an executive one, imposed upon the doctors and nurses. Respondents felt that these decisions served the institution first and the patient last.

In a parallel to social media and data companies such as Google and Facebook, these digital health record platforms generate data and enable data collection and trade on a massive scale. In the United States, electronic health records also provide the raw material for the subsequent production, management, and deployment of data assets on an immense scale. Neither the HITECH Act nor H.R. 4613 changes the fundamentally asymmetrical relationships of patient data. Simply because patients can access limited portions of their health records online does not mean they enjoy all the promised benefits, and at least one study showed that most patients do not access their data through online portals.[27] But if there is to be a revolution in healthcare to empower patients through their data, the healthcare lobby or the data companies that write legislation to solidify their control over patient data will not lead it. As long as there is no threat to the longstanding relationship between patient data ownership and the assetization of these massive collections, the promised revolution will not be digitized and will not be data-driven.

4 Mobilizing Alternative Data

Until recently, the term *psychographics* was little known outside Madison Avenue advertising agencies and marketing firms. Born in the marketing industry in the 1960s and 1970s, psychographics are the qualitative data on what makes individual consumers tick—the attitudes, values, sentiments, feelings, desires, and fears of ordinary people. It turns out millions of people may share one's seemingly unique personality, or at least aspects of it, and that personalities can be quantified, categorized, and grouped. In the early days of this marketing innovation, surveys, face-to-face interviews, and focus groups manually collected these data, which market researchers then analyzed with the early iterations of computer-based analytics and databases such as the marketing information system (MKIS) (Bradley 2007, 11). The focus on understanding the psychology of the individual consumer, and how to use those insights to drive behavior, marked a seismic shift for the marketing industry. Before the 1960s, marketing researchers focused on grouping people based on similar demographic characteristics—income, gender, race, religion, and where they lived—to target their communications en masse. They would combine the findings they gleaned from focus groups with facile understandings of human psychology to become what we now call psychographics. This analytical innovation shifted the focus of marketing persuasion techniques away from mass appeals to target the individual. Psychographics enabled marketers to segment millions of audience members into ever more refined categories, or segmentations, of consumers based on affinities, values, and emotional behaviors. Marketers use psychographics and segmentations in up to ten thousand marketing messages a day—through branding and advertising—across multiple platforms to tailor more targeted messaging, and to help advertisers cut through the noise to reach customers individually.[1] In the last four decades since its invention, psychographics has come to define how marketing is done.

Could an analysis of which of my friends' Facebook posts I have "liked," or what YouTube or Instagram videos I have watched, or even what size or color of jeans I have purchased on Amazon tell a marketing researcher somewhere out there the intimate details about who I am, what makes me tick, and how I might behave in the future? Not likely. The more probable scenario is that somewhere in the Nevada desert, machine-learning algorithms are churning away on a rented server in a data center that is capturing, inferring, and categorizing the digitized bits of my life into emotional, behavioral, demographic, and financial segments. How do the digital traces of my life, which these unseen and unknown others capture and analyze, empower algorithms and machines to predict what I might do, or to determine what my life chances could be? Or how, as philosopher Yuval Noah Harari has proposed, are algorithms empowered to know me better than I know myself (Harari 2017)?

The secretive uses of psychographics that have haunted our data were publicly exposed in March 2018 after whistleblower Christopher Wylie's explosive revelations about a little-known strategic communications consultancy: Cambridge Analytica. The company had weaponized the collective emotions, hates, desires, and fears of more than eighty-seven million American Facebook users for the direct benefit of the 2016 Trump presidential campaign.[2] When the Cambridge Analytica story first broke in spring 2018, it seemed to reconfirm a truth that many digital natives already live by: there is no privacy, and data is often used not to empower individuals, but to manipulate and control us.

The Cambridge Analytica story had been percolating in the US media since late 2015, when the *Guardian* reported that the company had acquired data from unwitting Facebook users to psychographically categorize and target voters based on their personality traits in order to help the presidential campaigns of Donald Trump and Texas Republican Senator Ted Cruz, all made possible with funding from ultraconservative Republican donor Robert Mercer.[3] Political campaigns traditionally use data to reach voters, and these data derive from public records, such as voter rolls and census tracts. Cambridge Analytica offered campaigns a competitive advantage through the unique "value proposition" of "powerful data-driven scientific methodologies" that the company claimed were developed by the "Behavioral Dynamics Institute, a leading international center for multidisciplinary research and development in behavioral change, including Target Audience Analysis (TAA), social influence and strategic communications."[4]

While Cambridge Analytica's strategic communication practices for right-leaning campaigns around the world had been garnering media attention since 2016, news of its nefarious data practices boiled over in spring 2018 with the whistleblower's revelation that the company had accessed the Facebook data of millions of users without their knowledge. The company had then used personality tests and statistical factor analyses to categorize targeted voters. Cambridge Analytica placed individuals, based on their purloined Facebook data, into one of the five temperament categories in the OCEAN personality model: *O*penness, *C*onscientiousness, *E*xtraversion, *A*greeableness, and *N*euroticism. The OCEAN taxonomy, or what is called the five-factor model, is considered the most validated and reliable model of personality traits used in psychology and neuroscience, but it tends to be understood more as a facile pop psychological explanation of personality, which market research often deploys (Allen and DeYoung 2017). Once Cambridge Analytica slotted the targets—the eighty-seven million individual Facebook users who had their data breached—into one of the five personality categories, the company then microtargeted them with political ads intended to appeal and to "speak" directly to whichever OCEAN personality Cambridge Analytica had identified for them. Through Facebook data purchased from Aleksandr Kogan, a psychology professor based at Cambridge University who collected the data from millions of unsuspecting users who took online "personality tests," Cambridge Analytica used the psychographic data and analytics in an effort to influence election outcomes in at least sixty-eight countries, including the Brexit referendum in the United Kingdom in 2016 and elections in Malaysia, Kenya, Trinidad and Tobago, and Iran.[5]

As a result of OCEAN psychographic segmentation techniques, combined with access to massive datasets that are cheap and easy to use, target and microtargeting dominate throughout the marketing industry. Marketers deploy these techniques across screen-based platforms, like phones or devices, and through older platforms such as direct mail (Bradley 2007). For instance, advertisers increasingly use a microtargeting technique called geofencing to deliver highly personalized messages. Sometimes called proximity marketing because the technique takes advantage of a prospective customer's proximity to tracking technology, geofencing allows digital marketers to send targeted advertising to a person's smartphone, like a digital billboard, by setting up an invisible boundary around a geographic location and accessing the location data on every phone that enters the parameter.[6] Use of geofencing can feel like a violation or breach of one's privacy, and especially in a doctor's office, which special privacy protections regulate, it can feel like a marketing intrusion into some of our most intimate and protected spaces. In early 2017, Copley Advertising planned to use the technology on behalf of an antichoice group based in the Boston area to send personalized anti-abortion messages, such as "You're not alone," Pregnancy help," and "You have choices," to the smartphones of young women who happened to be in proximity to reproductive healthcare clinics. Massachusetts Attorney General Maura Healey took legal action preemptively to prevent the advertising plan from going forward and cited the practice as an illegal use of consumer health data. Notably, the settlement was based on Massachusetts's consumer protection statute.[7]

The technique of dividing heterogeneous markets into tighter and narrower homogeneous groups of individuals with similar desires, attitudes, and sentiments in order to target them with tailored and personalized marketing communications have transformed how consumer goods and services are sold and consumed, and now drives virtually all commercial and political communications with the public (Doyle 2011). The use of data analytics that categorize individuals into particular profiles, fed by psychographic and other types of data, is not limited to marketing alone. Segmentation studies and psychographic data are used in electoral politics, policing, education, and public health research among many other sectors, as well as to feed predictive analytics and statistical model to help drive and shape behavior (Ebeling 2016; Grier and Bryant 2005; Mellet and Beauvisage 2019; Nielsen 2012; Slater and Flora 1991; Wu 2016; Zuboff 2019).

Often these psychographic data, as well as data that are not traditional for a given industry's use, are considered *alternative data*. In an interview, David, a data broker who works for a company that sells data analytics products to Wall Street investors, described the concept of alternative data as originating in finance: Hedge fund managers, in an effort to gain a competitive edge about a company to invest in, were looking for information that might provide new insights. David described how data exhaust—the data detritus sitting in a business's database—such as records of transactions with clients—can become productive alternative data that fuels the financial services industry and, increasingly, healthcare. David observed that businesses have "billions and billions of rows of data that they're not directly using in their day-to-day business. There's no real utility to them. What we do is partner with them, either through a revenue share agreement or through just directly writing them a very large check to get their data on a regular basis" (interview, May 20, 2015).

While alternative data was born on Wall Street, it lives in virtually every sector that participates in the data economy, from retailers to hospitals. In the preceding chapters I outlined just how much data are collected on individuals, above and beyond the digital footprints that we might leave online or through our wearable devices or phones. Millions of data points about our lives are collected, stored, and put to uses that are often hard to imagine because most data collection and uses are hidden from us. Our most intimate information is collected in spaces where we are likely to feel at our most vulnerable—in a doctor's office when receiving medical care, in a bank when applying for a loan, or even in a municipal social services office when signing up for housing or food assistance. These institutions demand of us an unearned and yet-to-be proven sense of trust at the point that our data are collected, a trust that we invest in a relationship or in a process to receive life-saving healthcare or life-changing credit. It seems fair that the information we hand over to a nurse or a loan officer should be completely relevant to us and to the reason we hand over the data in the first place: our data should be used to directly benefit us. A record of our blood pressure during a visit with a doctor, or the history of our car loan payments at the bank, all of this information, it is hoped, will be used in the future to guide how a clinician will treat us, or what kind of interest rate we can expect for a mortgage. Yet in a data-based society, where all sorts of digitized platforms (and the people who build or maintain them) are hungry for more and diverse data, this hoped-for outcome from sharing information with all of the institutions that we interact with, and possibly have little trust in, is far from the reality. Closer to the truth is that often our own data, and the data of millions of unknown others, can be weaponized against us.

At least two sociotechnical imperatives drive the hunger for data, which results in the data lakes and oceans developing in the healthcare sector. The first is machine learning and artificial intelligence initiatives, which need massive amounts of data to teach algorithms. The second is data platforms that build models that score individuals and that can predict risks and shape future behaviors. Shoshana Zuboff (2019) characterized the latter as a process of behavioral reinvestment. Focusing on information and analytics behemoth Google, Zuboff described how information brokers collect people's behavioral data—which in the hands of data analysts, such as Cambridge Analytica, can infer a lot about any individual's sentiments—to microtarget individuals with advertisements in a bid not only to predict their behavior, but also to drive consumer behavior to a desired outcome. She called this process the "behavioral reinvestment cycle," since these alternative data are used to modify consumer decision making, so that Big Tech can capture the surplus value of, and profit from, the changed consumer behavior (2019, 69–70). While Zuboff's analysis steadily focused on what she called the Big Other (in homage to Orwell's Big Brother)—the Googles and Facebooks of the world—much of the behavioral reinvestment cycle remains less visible because it occurs as part of business-to-business operations and other behind-the-scenes processes in places like hospitals or financial institutions. And these reinvestment cycles increasingly feed on alternative data.

ALGORITHMIC DETERMINANTS OF ALTERNATIVE HEALTH DATA

In preceding chapters I explored how in healthcare settings, such as the hospitals within the CMU Healthcare system, institutions collect, hold, and utilize patient information for purposes beyond immediate clinical care. These institutions transform a patient's private and identifiable health information into an asset that can produce future financial value. Within most healthcare settings, such as a doctor's office or a pharmacy, these data are primarily collected through a patient's electronic health record. Sociologist Amitai Etzioni observed that at the time that HIPAA passed in 1996, there was already an industry-wide drive toward healthcare systems adopting electronic health record platforms and networked databases to hold patient data. As early as the mid-1980s, some larger systems, such as the US Veterans Administration (VA), pioneered bespoke electronic health records platforms for the collection and dissemination of digital patient health data, which enabled easier health data surveillance for reasons other than individual or public health. Health insurers in the 1990s were already collecting nonclinical patient data as grounds to deny coverage, and third-party companies, such as credit bureaus and marketing firms, were linking lifestyle data collected to individual patients (Etzioni 1999, 142–43).

For healthcare institutions, alternative data can mean the information that supplements a patient's record and enhances clinical information concerning a patient's health status. These alternatives to clinical data are sometimes called a patient's "phenotypical data": information about income, housing status, and educational level. These data are broadly discussed as information that will help clinicians understand the social determinants of patients' health status or outcomes (Shilo, Rossman, and Segal 2020). These alternative data can help clinicians to understand in more nuanced ways how nonhealth factors affect a patient's clinical presentation. Larger healthcare systems, like CMU Healthcare, use these nontraditional data in what is called value-based care. Alternative data produce value for the healthcare institution in the form of either revenue or cost savings, or in higher ratings from creditors and from payers such as the Centers for Medicare and Medicaid Services (CMS).

The Affordable Care Act (2009) mandated the value-based care payment model for medical providers who receive CMS reimbursement, basing it on patients' health outcomes rather than per procedure performed. Because the provider has to demonstrate evidence of the patient's overall health for health insurers or CMS to pay for care, healthcare practitioners have turned toward data-driven solutions, algorithms, and data models. Nonclinical, alternative data feed these models, which help consider the entirety of the patient beyond their immediate clinical presentation or chronic healthcare needs. In theory, the value-based care model gives healthcare providers an incentive to deliver overall higher-quality medical care across the hospital or clinic, by paying them for reaching health outcome benchmarks. Rather than making payments in a "fee-for-service" model, based on the quantity of procedures provided to patients, the value-based care model pays practitioners for meeting targets on key indicators that demonstrate improved health. Codified by the CMS in the Medicare Improvements for Patients and Providers Act (MIPPA), the key indicators, such as rates of sepsis or patient readmissions after discharge from a healthcare facility, must be kept at certain overall target levels or providers will not be reimbursed for their services, and in some cases can be further penalized with the loss of federal funding.[8]

These value-based care incentives pressure healthcare providers to address structural inequalities outside of the clinic but are, nonetheless, seen as driving up healthcare costs. The turn toward understanding the larger sociopolitical structures that shape the health and wellbeing of an individual patient or an entire community encourages the never-ending pursuit of more data in healthcare. In many ways, this mirrors the credit reporting end of the industry; "alternative," or nonclinical, data complement "traditional" data and bring the data of a patient's electronic health record into sharper focus to build a more granular and refined understanding of the whole.

The inclusion of alternative data in analytics and models is seen as a key to unlocking systemic solutions in healthcare delivery and reimbursements, especially in the era of value-based care. If a patient comes to the emergency room for an asthma attack, a medical provider can treat the patient for the condition clinically. In the context of value-based care, however, they must also account for the social context that the patient lives in—those variables that more accurately predict whether the patient is at risk for readmission. Does the patient live near a power plant or a Superfund site? Does the patient have adequate and safe housing, free of pollutants that can aggravate asthma? Do they have adequate access to nutritious foods to strengthen their immune system? All of these variables now need to be accounted for in order to lower readmission rates, which can determine a healthcare facility's reimbursements from payers or eligibility to receive federal and state funding among other financial implications (Upadhyay, Stephenson, and Smith 2019). The incentives, of course, also help to reinforce the market-based model of healthcare.

Healthcare providers' CMS risk adjustment models, used to bill or set reimbursement rates from Medicaid and Medicare plans, do not include, for example, a variable for poverty to rate how well a hospital serves low-income patients (the CMS officially dubs these "safety net" hospitals).[9] These models rely solely on a hospital's patient readmission rates, without considering the social determinants in a patient's life that land them back in the emergency room. The risk model's inability to capture the larger social context, like the poverty rate of the hospital's neighborhood, that can drive readmission rates, puts pressure on healthcare providers to capture some of the social factors that impact a patient's health. Recognizing the direct financial impacts that it could realize by including alternative data in risk adjustment scores, health insurer United Healthcare used the heft of its market share to call for healthcare providers to include social determinants data in their diagnostic coding processes, specifically the codes in the International Classification of Diseases (ICD-10) clinical diagnostic Z category.[10] In the ICD-10, the Z clinical codes, specifically Z55–65, define social determinants—those social factors that pose a direct health risk and negatively impact health outcomes. For example, code Z55, along with its nine subcodes, categorized as "Problems related to education and literacy," can supplement a patient's health record to reflect the socioeconomic challenges a patient may face in finding a well-paying job that in turn may indicate poorer health outcomes. ICD-10 also includes a diagnostic, clinical code for homelessness, Z59.0 (Joynt Maddox et al. 2019). The Z60 set of codes and subcodes, which categorize a patient's experiences with social isolation, discrimination, and bias or what is labeled "acculturation difficulty," indicate how racism has negatively shaped the patient's health.

But it is hard to incentivize the extra coding that is asked of a healthcare workforce that is already overburdened with increased demands on the shrinking time given to face-to-face interactions with patients—many doctors are allowed only a ten-minute appointment window (Flaxman 2015). For a price, a healthcare provider can purchase social determinants of health data, such as the LexisNexis Socioeconomic Health Solutions data product sold to healthcare providers, to supplement the oceans of patient data that they already hold, and increasingly, the credit scores that consumer credit information bureaus provide.[11] These data are used as predictors of disease or proxies for other social determinant data, not necessarily to improve patients' health outcomes. Their purpose is to enhance a healthcare provider's risk adjustment scores and to realize more financial value (Dean and Nicholas 2018).

ALL DATA ARE CREDIT (AND HEALTH) DATA

The three largest credit bureaus make inferences about a person's health based on data contained within the bureaus' own financial information products. Customer car ownership, fast food purchases, medical debt, financing, and credit use for healthcare are types of information embedded within these data products (Wei et al. 2016). Analysts make inferences about a person's class or access to healthy food based on consumer and credit data, and these insights are sold to healthcare providers to make better sense of a patient's social context and what it might mean for their health. But this also means that a patient is no longer solely under the medical industry's gaze, but is made legible and subject to marketing and the debt society's surveillance (Lazzarato 2015; Waldby 1998). Without a credit history score, like the FICO score, or with a "bad" score, patients face further marginalization in healthcare settings because of their debt, purchasing behaviors, or even their financial and real assets.[12] For those whom the healthcare or banking industries already marginalize due to their immigration status, gender, race, or class, this profiling can have dire consequences for their health and negatively impact virtually all aspects of their lives.

In the hands of commercial interests, all data can be used to make inferences about health, even in the context of a covered entity that would be considered private. Patient health information produced in a doctor's office and protected under HIPAA can be exactly the same data collected from our smartphones but not protected under HIPAA. The privacy of data is thus context dependent (Nissenbaum 2011; Scholz 2016). Information that may be regulated for privacy in one setting is fair game in another (Schwartz 2009). Previously, with paper-based healthcare and records systems, these contexts were easier to control, ensuring that patient information produced in a doctor's office would not be (easily) re-identifiable in other, nonmedical contexts. But as data now swims in a vast digital ocean, these contextual boundaries have virtually disappeared.

The consumer profiles that credit bureaus hold will undoubtedly include data about an individual's health, types of medical services obtained, and the amount of medical debt that is in collections after 180 days of non-payment. For patients dealing with expensive medical diagnoses, such as cancer, these debts can reach the hundreds of thousands of dollars and will be reflected in the score, resulting in what is called medical financial toxicity. Since this information is about a patient's medical debt burden, it is still considered to be traditional financial information. But by utilizing "fringe alternative data," from consumer, social media, and web tracking sources, credit bureaus score customers through the application of behavioral predictive analytics, and they can make inferences about a person's health and medical information, either based on HIPAA-regulated information or through consumer data gleaned outside of HIPAA protections (Robinson + Yu 2014).

Since credit bureaus hold massive amounts of personal identifying and nonfinancial data on most Americans, virtually all the identifiers that constitute PHI overlap with data the credit reporting companies hold. Government agencies also use the credit bureaus to perform background and credit checks. For example, since the implementation of the health insurance exchanges under the Patient Protection and Affordable Care Act (ACA), Equifax has served as the financial verification vendor to the HHS for the health insurance exchanges.[13] This means that each time a patient enrolls on the health exchange, Equifax conducts a credit check and verifies income levels and other financial information on the enrollee.

With the passage of the HITECH Omnibus Rule, HHS intended to strengthen the privacy and security rules for patient data by directly regulating business associates, those entities that handle patients' PHI by providing services to covered entities.[14] Essentially, HIPAA regulates a subset of industries that use PHI rather than regulating all industries together. This is why the whistleblower working for Google's Project Nightingale project discussed previously was so concerned about the potential privacy vulnerability of fifty million Ascension patients and their protected health data. The boundaries between regulated data sectors are porous and expose health data to serious privacy risks.

Diverse data about individuals, or even their friends, neighbors, or complete strangers who share a psychographic categorical trait, end up in the databases of some of the most powerful information brokers in the United States: credit bureaus. Once in the hands of the credit bureaus, an individual's "traditional data" are combined with other data to feed scoring instruments for financial risk and consumer behavior, as well as "decisioning tools" that healthcare providers use to predict whether patients will pay their bill or take their meds, and to support clinical decision-making. Credit bureaus also sell these data to third parties that use health data for marketing purposes. Once private patient data leave the control of a covered entity, or the business associates and third parties that have direct contact with private patient information, these data repurposing activities occur outside HIPAA's regulatory protections.

The three largest credit reporting agencies in the United States—Experian, TransUnion, and Equifax—are, in fact, huge marketing and consumer information brokers, sometimes called data aggregators or data furnishers, because they are uniquely positioned to both collect and distribute financial and consumer data on virtually the entire population of the United States. Collectively, the three bureaus gather more than 4.5 billion bits of information each month to feed into their credit reporting files.[15] Experian alone claims to hold data on 98 percent of American households.[16] In total, these three companies potentially collect more data on the average American than any other information-based company that holds data assets on individuals, with each firm holding data profiles on more than two hundred million adults in the United States (Consumer Finance Protection Bureau 2015).

These companies rely on a broad collection of diverse data from public and private sources—including but not limited to financial information—that their predictive algorithms use to segment consumer behavior for targeted marketing, to determine credit scores (such as the FICO score), or, increasingly, to use for AI and machine learning (Consumer Finance Protection Bureau 2012). The FICO score encompasses the suite of credit-scoring instruments that Fair Isaac Corporation (FICO) offers, but as public awareness of credit scores has increased in general, many FICO products used specifically to score consumer credit, such as for a payment plan for a mobile phone, are referred to as "the FICO score" (Fair Isaac Corporation 2012). Additionally, other credit-scoring models, such as the VantageScore credit score, contribute to a credit risk score, stemming from a joint venture founded in 2006 by the three major credit bureaus. As a result, the industry continues to broaden and diversify what kinds of consumer data it collects, and once it collects them, how it repurposes and resells these data.[17]

The data profiles that these bureaus hold on individual consumers include information about an individual's credit use, such as a rental history, number of lines of credit held, and any history of late or missing payments. But since these companies are, in fact, data analytics and informatics companies, they also collect, process, analyze, and repackage nonfinancial or noncredit alternative data to create new data products. The credit information industry brings in four billion dollars annually in revenues from selling consumers' alternative, noncredit relevant information and other data-based products to third parties (Bennett 2017). Through data-use agreements and other contractual arrangements with more than ten thousand data "furnishers," Experian, TransUnion, and Equifax procure massive swaths of diverse consumer and public information about individuals, which they use to update the monthly files they maintain on customers. These furnished data derive from a variety of sources, from public records for registered deeds, marriage and divorce proceedings, voter registrations, and driver's licenses, to sources of "private" data, such as credit card transactions, online purchases and searches, social media usage, and metadata and locational data from mobile phones. Some furnishers have attracted controversy for collecting data on vulnerable people, such as rape survivors, individuals with a mental health or cancer diagnosis, and people who are HIV positive, and then selling that data onward to third parties (Dixon 2013). As long as one participates in the US economy through banking, making retail purchases, renting or purchasing a home, enrolling in the Transportation Security Administration's PreCheck program, or registering for public benefits such as subsidized housing or food assistance, at least one, if not all three, of the largest credit bureaus will collect, retain, and verify that individual's data (Hurley and Adebayo 2016, 157).

POROUS DATA

Everyday activities—from buying groceries to looking up interest rates on a mortgage website—leave a trail of data "exhaust," identifying data that will be scraped, traded, or sold to credit bureaus (Poon 2007, 2012). This information, though irrelevant to the consumers who inadvertently dropped these digital breadcrumbs, is highly lucrative to commercial data brokers.[18] These are individually identifiable data, originating from a variety of sources, including the metadata on smartphones, social media profiles, location data, and IP addresses; credit and debit card monthly statements; receipts and customer loyalty cards used at brick-and-mortar retailers; as well as patient health information. The data often significantly overlap with those individual identifiers, and companies can digitally identify and track consumers through the data exhaust produced in the data economy in myriad ways. Even when consumers attempt to go off the digital data grid, Big Data can still identify, track, and eventually categorize them—even as potential criminals.[19] In chapter 6 I discuss how those who don't participate in the formal economy, often those the banking industry calls "credit invisibles," are still visible through these data.

Contained within these data profiles, of course, are information that often overlap with the identifying data contained within other sectors that are under different data privacy regulations, including patients' PHI. Therefore, while a doctor's office may de-identify these data before sharing them, a patient's data are completely identifiable within a data broker's database, though ostensibly regulated by financial privacy laws, such as those established under the Gramm-Leach-Bliley Act, among others (Federal Trade Commission 2002).[20] The HIPAA Omnibus Rule expressly considers information such as a patient's name, date of birth, email address, home address, phone numbers, driver's license number, medical record numbers, and Social Security number, among others, to constitute the eighteen data points that make protected health information identifiable. When individually identifiable information is combined with a patient's physical or mental health condition for the provision of care or for payment for such care, these data fall under HIPAA PHI protections. However, HIPAA protects PHI only as long as it remains in the hands of a healthcare provider, health plan, employer, or healthcare clearinghouse.[21]

For health-related data to be mobilized outside of HIPAA's regulatory protections, a patient's PHI must be stripped from the record, essentially revoking its status as a special class of data object; it then becomes "nontoxic" data that can be reused, repurposed in tertiary research, or bought and sold for marketing or any number of purposes.[22] Pharmacies routinely de-identify and sell the protected health information contained within prescription data to health and medical data enterprises (Ebeling 2016; Tanner 2017). These datasets are then mined, analyzed, and repackaged into data products that are sold to third-party marketers or other businesses, often for purposes not directly related to a patient's healthcare or future health outcomes. For instance, the health data company IQVIA (formerly Quintiles and IMS Health) provides prescription data to pharmaceutical companies that, in turn, distribute these data products to their sales representatives, who can use prescription data in their "detailing visits" to doctors who prescribe a company's drug.[23] Physicians have been refusing visits from pharmaceutical sales representatives in recent years; and as a result, more drug companies rely on direct data marketing techniques to reach these increasingly resistant medical professionals.

Since the purpose of these companies is to identify individuals for their creditworthiness or to target them for marketing, these data are not de-identified and are used to match specific consumers with marketers or creditors (Hurley and Adebayo 2016, 148). As a result, credit bureaus have direct and complicated relationships to patients' PHI. Credit bureaus' use or disclosure of PHI *should be* HIPAA compliant, but in practice this is impossible. Owing to this loophole in data privacy legislation, health data, health-related data, and data concerning health that are available from other sources are not subject to HIPAA, and therefore can be used or disclosed for different purposes. Thus, HIPAA regulates the information in the database of a covered entity, while the same information in the database of a credit bureau may be further disseminated. Possible negative implications for patients include having their health status disclosed through consumer health data, such as when a social media app discloses its users' HIV status to third parties, regardless of whether the data was produced outside their doctor's office. In 2018, for instance, it was widely reported that the dating app Grindr shared user data, including users' profile information about HIV status, with two third-party vendors. The company defended the breach with claims that it was "industry practice."[24] The exposure can feel the same: a person *feels* like their health privacy has been violated when they are denied credit because of their health status or they become a marketing target because of a medical condition.

DATA BREACHES AND THEIR CONSEQUENCES

"Protect the Data that Defines You, Mary," reads the subject line of the targeted marketing email I received from Experian. It is the latest marketing spam that I've received from the credit-reporting giant, trying to sell its "credit file monitoring" services back to its data breach victims. In the years prior to receiving this particular email, I had signed up with Experian's credit monitoring service after numerous data breaches had occurred with my health insurer as well as with Experian itself. This email illustrates how the credit-reporting industry puts the onus of data stewardship and protecting data from risk back onto the subjects of data.

A data breach underlines the potential risk and harm that can come from a credit information company that has custody of and access to identifiable health information without accountability. One example was when identifiable personal data, such as millions of customers' Social Security numbers and bank account details, which Experian owned and sold to a legitimate third-party, was exposed to criminal hackers.[25] While Experian was acting as a "good steward" of personal data by complying with data privacy regulations, this did not prevent a data security risk. In contrast, the Equifax data breach in 2017 demonstrated complete negligence of the company's duty to be good stewards.

In a public statement in September 2017, Equifax revealed that several months earlier hackers had illegally accessed the private and identifiable data of more than 143 million customers, about half the US population. While not the largest breach in U.S. history, the Equifax breach became a huge media spectacle, in part because of the egregious way that Equifax handled the fallout. In a press release dated September 7, 2017, Robert Smith, the chief executive officer of Equifax Inc. at the time, revealed that the company had detected as early as March of that year that criminals had exploited a vulnerability on one of the company's US websites and gained unauthorized access to "certain files" within Equifax's databases. After a company investigation of the breach, corporate officials determined that hackers had gained access to the names, Social Security numbers, addresses, birth dates, and in some cases driver's licenses of more than 143 million customers in the United States alone. Much of that information comprises the eighteen data points that can be used to identify an individual. Moreover, the alleged hackers stole the credit card details of 209,000 US consumers and the personally identifiable information of another 182,000. Because the financial data company operates internationally, it also determined that some customers in the United Kingdom and Canada had had their records compromised.[26]

Any illegal access to personal data that a third party, such as a credit reporting agency, controls is very concerning and can have profoundly negative consequences for consumers. Where other data companies had already decentralized consumer data and dispersed information across several databases, so that if one database were hacked, the hackers would have only partial information on any given individual, Equifax still held data on consumers in one central database, making private data that much more vulnerable.

Equifax's failure to act more swiftly to protect consumer data eroded the little public trust that consumers and creditors had in one of the largest financial data companies in the world. Along with the perceived corrupt actions top management took (three senior managers sold some of their Equifax stock after the company discovered the unauthorized access), this negligent behavior severely damaged Equifax's reputation, and it led to a Department of Justice criminal probe against the company. The Equifax hack was considered the worst breach of private data in recent US history, not because it was the largest, but because of the very sensitive nature of the compromised information (Villas-Boas and Pelisson 2017).[27]

In the specific case of the Equifax breach, the compromised data also correlated to the private data of millions of patients, which can be linked to health data regulated under HIPAA. While the media gave most of its attention to the highly sensitive financial and personally identifiable data of hundreds of millions of Americans, they in large part ignored the fact that the breach could also put protected health data at risk. Many started to question how one company could have so much power over consumer data without accountability. What was less understood was how the exposure of these data could lead to dire consequences for the healthcare sector and the potential for direct harm to patients.

The financial data and credit-reporting industry owns and controls much of Americans' private health data; as we have seen, these can be the same data that, when held by covered entities, is subject to HIPAA regulations. When these health data are held by credit bureaus, they are no longer HIPAA-protected. Once data are mobilized outside of the HIPAA regulatory context, primarily through de-identification, or if the data are collected in contexts outside of healthcare, say from online or retail transactions, the data are not subject to HIPAA and can be utilized for myriad purposes. These purposes can be to patients' detriment, instead of furthering their health and well-being. The inevitable exposure of intimate and sensitive personal data handled by third parties, such as credit bureaus, can be attributed to a range of factors, from unknowing noncompliance to malfeasance and criminal negligence, or even to just doing business as usual. But the risks of mishandling and exposures are always high because of the fragmentary nature of data privacy laws in the United States, where data are regulated by sector. In contrast, the European Union's General Data Protection Regulations (the GDPR) handle data regulations through omnibus legislation.[28]

Credit bureaus not only repurpose patient data, they also increasingly adapt behavioral and consumer predictive analytics in order to create data packages that they sell to healthcare providers as tools to predict patient behaviors, such as medicine adherence or the probability of readmission. Thanks to a loophole in the HIPAA regulations, credit bureaus and other data analytics firms are held to a far different standard than the entities that collect the patients' data. HIPAA regulations charge healthcare practitioners with ensuring patient data confidentiality and security. Data stewardship and data regulatory compliance are often seemingly impossible tasks, yet healthcare practitioners are subject to substantial penalties when something goes awry. By contrast, the credit bureaus probe, prod, dissect, and combine health data surreptitiously collected from us—sometimes with our consent, but most often not—through our interactions with the healthcare system and our daily routines. Credit bureaus' data ownership and dissemination practices underline to what extent individuals (as patients and as consumers) are subjects of data—and, as I discuss in the next two chapters, subjects of debt.

5 On Scoring Life

Measure what can be measured, and make measurable what cannot be measured.

—Popularly attributed to Galileo Galilei

Personalize, predict, and prevent. Since the first human genome was successfully sequenced in 2003, the field of genomics has accelerated and unleashed a new marketing hype-cycle around "precision medicine." The promise of precision medicine, proponents say, is that the use of a patient's own genetic information can, when integrated into statistical modeling, machine learning, and artificial intelligence systems, help to predict adverse health outcomes. Based on this data, medical providers can customize and personalize their clinical care to an individual patient, in the hopes of mitigating risk (McDonald and Murphy 2015). Precision medicine's promise is based on the presumed predictive power of risk scores that rely largely on massive datasets that have little to do with the individual patient. So, while this promise of customized healthcare based on one's own genetic data remains largely unfulfilled for most patients, the drive to develop predictive scores for patients has not slowed down.

From the moment that we take our first breath, we are quantified—given a score. Our data throughout our existence, be they biomedical, financial, or psychographic, are captured from a variety of contexts, from health records to Facebook accounts, and they are destined to be reborn in new and varied contexts. Once these data are captured, they're stretched beyond the boundaries of our individual lives, where they join an amalgam of billions of data points and are made to perform "work": data analytics is about making inferences out of the aggregate. Some of that work is in predicting the future, where our data are instrumentalized into scores (Citron and Pasquale 2014). These scores can function as a tool to support a decision made about us, or to spur us to make a change in our own lives that may directly increase our well-being. In other instances, scores are used to augur or predict something about us, like how risky our behavior or financial situation is or may become. Scores can help to visualize and make graphic the past, present, and future of an individual's life. Scores are also used to manage and anticipate what might happen, good or bad. When a situation receives an objective number, action can be taken. In the words of Carl, the data scientist working for the Consortium introduced in chapter 2, "If it can be measured, it can be managed. We really say you can't change what you can't measure."[1]

Scores make the quantification of risk visible, and quantifying risk gives one the sense that the uncertainty of life can be knowable and controllable. But life is often unmeasurable. Making it quantifiable and subjected to scoring has become both the process and the currency in data-based and AI-supported decision-making (Porter 1995). Our society is utterly in the thrall of mundane datafication and calculative automation (Mayer-Schönberger and Cukier 2012; Woolgar and Neyland 2014). The numbers derived in these processes have great power in determining life chances or outcomes, but the processes themselves are far from certain, transparent, or fair, and they often occur with minimal intimate knowledge or trust in the object—us.

When I gave birth to my child in the mid-1990s, she was placed on my naked stomach within seconds of her birth for skin-to-skin contact, only to be whisked away from me moments later to be weighed, measured, and given an Apgar score at minutes one and five of her life. The Apgar score rates on a zero-to-two scale a baby's skin color, reflexes, heart rate, muscle tone, and efforts to breathe, to calculate the baby's overall score. The Apgar score can help a clinician determine if the baby had any brain trauma from the birth, any signs of infant jaundice, and may help to detect very early signs of other neurological problems. A baby's Apgar score, along with all of the other measures performed within her first five minutes of life, will be entered into an electronic health record on a networked computer close to the bedside in the labor room. The Apgar score will be used only to give a pediatrician an assessment of the baby's health within minutes of birth. It is not designed to predict a baby's long-term developmental outcomes and will not follow a baby in her health record for much longer after going home.

There is an ebb and flow to the relevance of scores given to us throughout our lives. While the Apgar score might be used clinically for only the first hours or weeks of a baby's life, as the child grows, without a doubt she will continue to be measured. Her body produces new information—data—that will be quantified and modeled, and she will be given more scores. Her scoring will certainly be based on her health—as long as she continues to have access to healthcare—but in a highly data-centric society, she will also be scored throughout all aspects of her life, and these scores will be repurposed for uses well beyond her individual life. She will be scored at school, where standardized tests and grades will be used as predictive scores not just about her own chances of academic success, but also about those of her classmates and millions of other students. Data will be collected about her and used to score her on where she socializes, on where she shops, on how mobile she is, on whether she should be trusted with credit, or how likely she will be to organize or join a union at her workplace.[2] As long as she interacts with institutions, or any center of calculation that uses measurement and data to manage risk or to predict future outcomes, she will be scored. These scores can be used to improve her life or they can haunt her, branding and stigmatizing her.

SCORING THE PAST TO UNDERSTAND THE PRESENT

Other scores are used to predict the risks of developing conditions, or to determine the life chances for the subject based on the past. The Adverse Childhood Events (ACE) score is an example of such a score. The ACE score quantifies one's exposure to childhood trauma to predict resiliency throughout the life course and to calculate the risk of developing chronic disease or drug dependencies later in life. The score is on a one-to-ten scale, with a score of zero representing no adverse childhood experiences, and any risk score higher than four indicating that the person is likely to experience serious and multiple health and behavioral problems in adulthood. The higher one's ACE score, the higher one's probability for heart or lung disease, drug or alcohol abuse disorders, a sexually transmitted disease, or depression or a suicide attempt, among other chronic diseases or social and behavioral problems. An ACE score of four, for example, assesses one's risk for a suicide attempt (twelve times higher than an ACE score of zero) and for developing bronchitis (four times higher than a zero score); an ACE score of six predicts that one's life will be shortened by twenty years (Brown et al. 2009).[3]

The CDC and insurer Kaiser Permanente conducted the studies that contributed to the development of the ACE scale. The HMO surveyed seventeen thousand of its members in Southern California in the mid-1990s with questions regarding childhood experiences and their current health behaviors and status (Felitti et al. 1998). The ACE score has performed important work in making visible how early toxic stress can profoundly shape a person's risk for developing chronic disease or poor health and social outcomes later in life. The score has helped to shift stigma and blame for conditions like substance abuse away from the nebulous "culture," or a patient's personal failing, and toward a larger narrative of trauma and systemic deprivation, such as poverty or structural violence.

Since its invention in the late 1990s, the ACE score has come to be used more broadly in health and social support settings, where practitioners and counselors look to a client's score to inform the care of patients with drug-dependency disorders, educators consider it in using trauma-informed pedagogies with their students, and it informs support offered to formerly incarcerated people transitioning through parole and into life after imprisonment. As part of what is called "trauma-informed care," social workers and public health researchers increasingly use the ACE score with vulnerable groups—such as clients who register for publicly funded treatment for substance abuse or parolees living in transitional housing. Potentially, counselors who work to support formerly incarcerated people reentering their communities can use the ACE score as one of many bits of data to inform their care and to provide more holistic, client-centered "wrap-around" services.

Whether a patient or client even has an ACE score may say a lot. The chance that an individual with stable socioeconomic resources and strong social ties will receive an ACE score in social services settings is rather slim, leaving those who are less stable or able to access the resources to survive and thrive to be quantifiably rendered and marked as risky subjects. The ACE score isn't designed to include variables that increase resilience in a person's life, such as a caring teacher or mentor who enabled them to survive trauma, or strategies that helped them to be resilient, or supportive relationships or therapy in adulthood. The ACE score helps to objectify one's past trauma so that its ongoing intrusions in the present can be managed or, at the very least, better understood.

The ACE, like other scores, can also label or stigmatize people. Beyond describing how early trauma impacts one's present life and perhaps serving as an indicator for more social services support, there is a danger that the score could be used in deterministic ways. Judges and parole boards have integrated the ACE score into assessments and algorithms that predict the risk of recidivism among youth offenders, and to support decision-making in cases. Rather than using the score to inform care, the criminal justice system uses it instrumentally and punitively, with the potential to stigmatize young parolees for years beyond when they have served their sentences. Some early investigations show that a young parolee's ACE score does not even reliably predict future recidivism (Craig et al. 2020). Using the ACE score to predict the likelihood of a future offense, rather than to guide support services postrelease in order to ensure successful reintegration into the community, distorts the score's purpose. The inclusion of ACES in such risk-scoring instruments has the potential of further traumatizing survivors of early childhood trauma, and it defeats the score's intended purpose of making one's pain and suffering visible through quantification in order to understand better a survivor's current situation and to provide trauma-informed support.

RISKY SUBJECTS OF TRAUMA

The data-sharing Consortium introduced in chapter 2 held a workshop at CMU to bring together stakeholders on the benefits, pitfalls, and future of the cross-sector sharing of public health data. The workshop had more than two hundred attendees, most either data managers from local government agencies or social workers from nonprofit social support services organizations, such as the United Way and smaller charities that bridge the gaps in essential social services due to government cuts and austerity measures. Attendees exhibited some hesitation about exactly what kinds of client data they should share with the Consortium. The participants were concerned that sharing certain types of very personal and potentially painful information would produce unintended harm.

Many who work in social services, such as Zeynep, a housing specialist for a suburban county government department who attended the Consortium's workshop, are keenly aware of the danger that these data-driven platforms will reproduce biases against their clients (Eubanks 2018). Zeynep, a geographer by training, oversees a swath of housing issues for the county in addition to providing "coordinated reentry" for constituents who are at risk of homelessness or who are already living without stable housing. Zeynep came to the workshop to learn how to better use data—personal data on residents that the county already holds as well as aggregated data collected from a variety of third-party sources—in a coordinated effort across departments to improve access to county-provided social services. Thinking of her clients as she listened to the presentation on the Consortium's aims, she saw some red flags about what information should be shared or included, such as risk scores used for homelessness and the ACE score, which may not tell the whole story of a person's life.

In particular, Zeynep's concern was about the use of segmented, aggregate data to "make sense" of a client's struggles with the cascading health and social challenges that produce housing insecurity. The county's data system platform, called Case Worthy, uses these data to enhance a client's file and to coordinate care across all the social services departments involved in coordinated reentry. Zeynep explained that these segmented data often come from building and flood insurance companies, but sometimes the origins of these data are unclear, and as such, the data have a way of obscuring rather than elucidating the complexities of each client's unique situation. Her trust in these data to tell a client's story is low: "So it depends on which group you are looking at, and you see it's really working on the most needy [clients]. Given the data related to these segments, I know, since I am looking at the... social indicators [data], and I don't usually rely on the segmented data there.... I don't trust this, because the margin of error is so high." Zeynep knew that data, especially unreliable data that don't provide a caseworker a nuanced story of the person in crisis, can "mark" clients and follow them for years. She considered a possible scenario: "One day, you might have a job, the next day you might be laid off, and then suddenly you find yourself at risk of homelessness." If you call the county homeless hotline, you will hand over your information, with a level of trust that the information will be used to help you in your crisis. A case file will be created for you, and your data will be subject to assessments and analysis as to whether you qualify for resources, and at what level. All of this will be recorded in your file. "You are trusting us that [your] information will stay with us forever," Zeynep observed, and she asked, "Maybe you don't want this?" Zeynep worried that these assessments will follow people despite changes in their circumstances.

Many working in social services, like Zeynep, are also concerned about biases built into these data, scores, and algorithmic systems, which rely on machine learning and artificial intelligence for their automated analysis. Another workshop participant expressed discomfort at the thought that her clients' ACE scores might end up in the regional data hub. She was uneasy that an ACE score would follow her clients, who had little control over how their data are used, and possibly retraumatize clients already living in precarious conditions. She was also concerned that the score would further stigmatize her clients from marginalized backgrounds, and in particular her Black and brown clients, and potentially cause punitive harm.

There is a danger that scores, especially those that predict risk, when they are automated or used as part of a model, will reproduce traumatizing social biases against stigmatized groups in particular (Benjamin 2019; Obermeyer et al. 2019; Skeem and Lowenkamp 2016). A study of some of the most widely used predictive algorithms that support clinical decision-making in assessing a patient's risk for an adverse health outcome, such as the Vaginal Birth after Cesarean (VBAC) score, the glomerular filtration rate (eGFR) score for kidney function, or the American Heart Association's (AHA) Heart Failure score, found that many predictive algorithms replicate existing racial biases in medicine, and can have the potential of denying patients needed medical care (Vyas, Eisenstein, and Jones 2020; Williams, Hogan, and Ingelfinger 2021). Extensive sociological investigations into the scores that the criminal justice system widely uses to determine the risk of reoffending reveal that many of these scores are racially biased. These scores include those within areas from policing to parole, such as the widely used Correctional Offender Management Profiling for Alternative Sanctions (COMPAS) (Garrett and Monahan 2019). In the case of COMPAS, the score is a combination of the incarcerated person's history of criminal offenses and "two dozen 'criminogenic needs' related to the major theories of criminality, including 'criminal personality,' 'social isolation,' 'substance abuse,' and 'residence/stability.'"[4] Many of these factors align with data connected to the social determinants of health, which increasingly contribute to predictive risk scores that healthcare and other areas of social services use. The COMPAS tool—which some jurisdictions adopted early on, before adequate and independent validation testing could be done—produced risk scores that were much lower for white defendants compared to Black defendants, even for those with no prior history of offense.[5] In one widely cited assessment of the COMPAS tool as used to determine sentencing in Broward County's (Florida) criminal courts, the higher the risk score that a convicted felon received, the longer their sentence; the higher risk scores skewed against Black defendants (Rudin, Wang, and Coker 2020).

Many states have made the use of these recidivism risk scores advisory, rather than mandatory, in sentencing decisions. Sociologists Hany Farid and Julia Dressel show that the algorithms used in the prediction of criminal recidivism (which are rooted in the belief that algorithms and data-as-evidence can solve deeply embedded structural problems) are no more accurate or unbiased than the predictions people with little or no expertise in criminal justice make (Dressel and Farid 2018; Farid 2018). Given that the US criminal justice system overpolices and criminalizes Black and other communities of color at higher rates than other demographic groups, how useful is an algorithmic tool that purports to predict an individual's chances of reoffending? The overdependency on algorithmic risk-scoring tools in high-stakes decision-making, such as around sentencing or parole, was never the intended use of these predictive scores; they were designed to inform the process but not to be the sole determination of the process (Chouldechova 2020). Such tools perhaps are better at measuring the criminal justice system's biases, or one's chance of being targeted by police or the courts for rearrest and imprisonment, than the probability of engaging in future criminal behaviors (Skeem and Lowenkamp 2016).

As with recidivism risk scores, overreliance on psychological or "personality"-based risk scores to determine whether a client is in danger or in need of support can be stigmatizing (Courtland 2018; Eubanks 2018; Madden et al. 2017). The Allegheny Family Screening Tool (AFST), which that Pennsylvania county's department of human services uses, is one such predictive risk score developed to help caseworkers identify children in the family services system who are at risk of abuse or neglect. Initially implemented without adequate calibration measures to address possible biases, the tool was based on scant data inputs. Rolled out at a time of significant changes in statewide child welfare laws, which were not integrated into the tool, the algorithm did not account for changes in a client's life, such as recovery from a drug abuse disorder or improvement following therapy (Eubanks 2018, 127–73).[6] The algorithm also limited a social worker's ability to override the score; at times, they adjusted their risk assessments to match the score the AFST tool suggested (Zerilli et al. 2019). In essence, rather than supporting the human beings making decisions in the evaluative process as intended, the AFST score became an AI-adjudicated determination that potentially could tear families apart (Chouldechova et al. 2018, 14; Courtland 2018). After the AFST scoring tool garnered much public scrutiny and criticism, informatics researcher Alexandra Chouldechova and her coauthors (2018) reanalyzed the tool for bias. Through their research they found additional problems in the instrument's original validation—the process of determining the accuracy and reliability of an algorithm—and in its ability to predict risk reliably (Chouldechova et al. 2018, 5). Many of these machine-learning and artificial intelligence platforms also stigmatize the poor by the process of selection, since most of those who can afford to pay for mental health services or drug rehab will remain off the social services radar and thus invisible to these predictive risk platforms. Individuals who cannot afford to be invisible or to protect their privacy from these platforms become hypervisible, oversurveilled, and overpoliced (Bridges 2011; Madden et al. 2017). Scores can be used to obscure, hide, or unduly amplify certain kinds of risks, and they have the potential to stigmatize or penalize people who are supposedly more prone to particular risks than others. This tendency is a shared concern for those who build the models that create the scores.

MODELING LIFE TO CONTROL RISK

In his one-paragraph short story "On Exactitude in Science," Argentinian poet and author Jorge Luis Borges articulated a complex thought experiment folded into a sly critique of colonial power and empire, all with the precision of poetry:

> In that Empire, the Art of Cartography attained such Perfection that the map of a single Province occupied the entirety of a City, and the map of the Empire, the en-tirety of a Province. In time, those Unconscionable Maps no longer satisfied, and the Cartographers Guilds struck a Map of the Empire whose size was that of the Empire, and which coincided point for point with it. The following Generations, who were not so fond of the Study of Cartography as their Forebears had been, saw that that vast Map was Useless... [i]n the Deserts of the West, still today, there are Tattered Ruins of that Map....[7]

In their attempt to construct a model of the Empire—an exact one-for-one abstraction that could account for every inch of the land—Borges's cartographers built into the map its own failure. The map of the territory is not the territory. Or in the case of the algorithmic models used to score and predict a person's chance of developing heart disease or defaulting on a mortgage, these models are imperfect or incomplete analogues to lived experiences. It is impossible to account for all the variables that need to be included for the model to be accurate. So while all models are wrong, some can be useful in their approximations, in that they provide a mental map or a conceptual frame for the object of study (Box 1979, 792). Scores are impersonal and imperfect approximations of an individual's risk to develop a condition or encounter a situation (for good or bad, usually bad) because the algorithms that make the scores are based on massive datasets collected from millions of other people in a variety of culturally and socially grounded contexts.

The process of making data fit a model rather than a model reflect the data became clearer to me one evening, after most researchers had gone home, when Theresa and I were working alone in the lab. Earlier in the day, I had overheard a conversation in the cubicle next to mine between Dr. Jim, as he is affectionately known among the lab members, and Jinling, the postdoc working in Theresa's lab. I had puzzled over it since then, so I took advantage of the quiet evening to ask Theresa to help me make sense of it.

Dr. Jim, a cardiothoracic surgeon, had walked across the hospital campus to meet with Jinling, who was helping him to build a model—a predictive risk tool—that would help identify whether a patient was a good candidate for surgery or if her condition could be better treated in other ways. In his early sixties, Dr. Jim was a veteran surgeon who had worked at the CMU hospital system for decades, but he was also a keen medical informatics researcher who wanted to use data as another surgical tool: in this case, as an instrument to predict risk and to aid in making clinical decisions.

After looking at some of the dataset that Jinling had used to build the model, Dr. Jim became audibly frustrated. "[This] garbage data comes in and makes our model useless, Jinling," he said. "We need a 90 percent accuracy from the model to predict chances [to recommend heart surgery]. You need to throw out data that doesn't fit, that doesn't make the model work." I could hear Jinling agree that she would rerun the model again without the misfit data.

This whole exchange seemed puzzling, especially since Theresa and I had previously discussed what kinds of clinical data go into risk modeling. I asked her if she knew what Dr. Jim meant by "garbage data." She knew exactly what he meant. "He is using a complete case analysis in order to build this predictive model," Theresa explained. She pulled out a blank piece of paper and drew a "U" on it, and then drew a series of lines underneath the U. Pointing to the series of lines with her pen she explained that within a dataset of heart disease patients at the hospital, for example, every patient's record should "match" before the dataset was used in the model. No case's spreadsheet should have any empty cells: they should all have a value, some information in them. Some data scientists, she explained, are concerned more with convenience sampling in order to fit a model, rather than with making the model reflect an epidemiological phenomenon. "They work to make data fit the model, tossing out what doesn't fit or what is missing," she said. All the data lines need to be the same for the analysis to work. "Convenience, for the data scientist," she explained, "is to say that all those tossed cases are 'missing at random.'" Of course, there is nothing random about which cases are missing. While she understood that for data scientists, the goal is to build a validated and elegant model, Theresa saw tossing out data as building a selection bias into the model.

As an epidemiologist first and a data scientist second, Theresa worried that the cases that get "tossed out" are patients who need medical attention, or worse, patients who could die by being "tossed out" of the model. Because their records are incomplete, the patients themselves, not only their data, become the detritus that gets tossed out of the model. Turning to the U that she's drawn, Theresa showed me where those "missing at random" patients might be tossed out:

> Suppose the model is a U-shape graph with both ends of the U populated by patients who are well off and can afford high levels of care—so their [electronic health] record has a lot of data. Or they are very sick, so their records also have a lot of data, so they will be on the other end of the U. But what about all of the patients in the trough of the U, at the bottom of the U? Those are the missing patients, patients who were tossed out because they didn't have enough data but who may need attention and care regardless.

In his book *The Everyday Life of an Algorithm* (2019), Daniel Neyland described how data scientists make data fit so that a model will "work." During his fieldwork, Neyland observed the development of an experimental machine-learning algorithm for use in airport surveillance systems to visually distinguish between "safe objects," such as luggage moving on a conveyor belt or being carried by a passenger, and "unsafe objects," such as a lone suitcase that might be abandoned in an "out-of-place" location. The goal for the scientists building the model was to take video surveillance footage and teach the algorithm to identify objects and distinguish between "luggage-shaped objects" and "human-shaped objects" that move through the secured and unsecured areas of an airport. If successful, the algorithm would flag those objects that were "out of place" as a danger. During one development meeting, Neyland observed the computer scientists speculate about what it would take for the algorithm to be deployed in the real, and messily random, setting of an airport, and still work as intended. They joked that it was simply a matter of raising the heights of ceilings, or narrowing passageways to control the flow or gaits of passengers (Neyland 2019, 35). For the model to work perfectly, the real world had to be made to fit the algorithm, rather than the algorithm built to reflect the in situ context of the airport.

Theresa sees a prime example of this problem in the Framingham Risk Score (FRS), officially called the Atherosclerotic Cardiovascular Disease Risk Algorithm (ASCVD Risk Score), or, as it is more widely known, the Ten-Year Risk Score. Theresa has built her career on the examination of the Framingham Heart Study (FHS), a longitudinal study that seeks to understand the statistical prevalence of cardiovascular disease among large populations. The FRS, which emerged from the study, calculates the chances of a patient developing heart disease over ten years. But, she explained, "Your chance or my chance of developing heart disease in ten years is really 0 percent or 100 percent." The FRS is a predictive score, based on a statistical probability, and thus cannot account for all the variables that can delay or prevent the onset of cardiovascular disease, such as a patient's ongoing access to health insurance or medical care, or healthy foods, or safe places to walk and exercise.

Commencing in 1948 with three thousand residents of a small city just west of Boston, the Framingham Heart Study now includes four generations of the original participants and became the gold standard in clinical care of cardiovascular disease. The Framingham Heart Study produced the phrase and concept of a "risk factor," as well as the instrument that uses risk factors as inputs, the Ten-Year Risk Score, which is used to help predict a patient's chance of developing cardiovascular disease in a ten-year timespan. The problem that Theresa sees in the Ten-Year Risk Score, as well as in the study that produced it, is the impossibility of a model including all the variables that contribute to the probability of developing a disease, not to mention the selection biases embedded in the study and the score. The Ten-Year Risk Score and similar measures are an imperfect analog to a patient's actual risk of developing heart disease because the score is based solely on biological measures, which are measured against the original Framingham study cohort. Those patients were white, generally more affluent as compared with the rest of the United States, self-selecting, and regionally confined.

The Ten-Year Risk Score and the Framingham study were both the gold standard in clinical care for cardiovascular disease until about 2015, when biomedical informatics researchers turned to more inclusive studies, both of data and of patients, to build models and construct a new algorithm that would more accurately predict risk for more patients. The Jackson Heart Study (JHS) was implemented in 2000 in Jackson, Mississippi, and like the FHS, is a longitudinal study. But that is the point where the two studies, and the scores, diverge. The Jackson Heart Study is community-based, intergenerational, and focused on Black Americans, and it includes biomedical, genetic, and social determinants of health data. To build a model that is closer to Borges's impossible map of the territory, biomedical informatics researchers used the combined data from the JHS and other cardiovascular disease studies, and recoded and revalidated the Ten-Year Risk Score. By fine-tuning and enhancing the original score with more and inclusive data—or the missing patients that Theresa worried about—researchers developed a risk-scoring instrument, the Pooled Cohort Equations. The new scoring algorithm provides a more individualized and less biased prediction of a patient's chance over ten years for having a heart attack or stroke (Schiros, Denney, and Gupta 2015). The Pooled Cohort Equations ASCVD Risk Score means that, for instance, the doctor of a Black woman aged fifty, who is not on statins and regularly exercises, can turn to the ASCVD score to provide a more accurate interpretation of the patient's risk of a heart attack. In turn, the doctor of a white man with the exact same risk factors might use the Framingham Risk Score. While the scores have been revised in an attempt to eliminate the algorithmic model's embedded biases, they are still not a one-for-one map of the territory—in this case the patient's body and future risks. Scores like these will circulate out of an individual electronic health record and into the larger lakes of data to perform new calculative tasks.

Risk scores connected to a patient's health do not only help the patient or her doctor to make a plan for her health. Hospitals and health insurers also use these scores to understand how much a patient, in the near and far term, might cost the healthcare system. Furthermore, health insurers buy directly from consumer bureaus what are marketed as "social determinants of health" data but are essentially lifestyle and consumer data. Yet these scores are used to determine risk levels of insured patients and to raise premiums on insurance subscribers as if they are objective measures of risky patients.[8]

SCORING THE PAST TO DRIVE FUTURE BEHAVIOR

Predictive analytics are based in statistical and computational science, machine learning, and artificial intelligence, all processes that analyze historic data to make predictions about future events or trends that are otherwise unknown. Wendy Chun has argued that predictive models and the risk scores that are derived from these models are best at "predicting the past," because such models utilize old data from millions of others, and they also rely upon old models of categorization (Chun et al. 2011).

The data and models that healthcare and medical settings use to build risk scores are no different. These risk scores rely on historic and imperfect data to predict someone's future chances of a health event. To improve these models and scores, healthcare providers turn to financial, social, and other kinds of "alternative" information—such as claims data that are sent to health insurers. Since credit bureaus and other third-party companies hold these categories of data on close to 98 percent of American households, these companies leverage and enhance the data to provide risk-scoring models to healthcare. The LexisNexis Risk Solutions product includes "lifestyle" data that the company sells to healthcare and insurers to help set premiums and score patients' financial risk levels.[9] How is it that the FICO score, the almost-ubiquitous credit risk score associated with one's financial health, has become an instrument that healthcare utilizes to make predictions about one's physical well-being?

Crossing Sectors to Score Risks

As the previous chapter discussed, the credit risk information brokers, such as Experian, collect and integrate "alternative data"—behavioral and psychographic information about individuals—into the segmented and consumer files that they maintain. Credit bureaus increasingly use consumer, behavioral, social media, and other data scraped from online uses to provide more fine-grained predictive analytics on credit risk, as well as to provide information to healthcare providers about a patient's large debt loads on credit cards, and other financing used to pay for medical services (Kiviat 2019). These scores are used not only to predict behavior, but also to nudge or drive future behavior.

Biostatisticians and public health researchers widely acknowledge that one's zip code more accurately indicates one's health than does one's genetic code. It is just another way of saying that the most obvious social influences, or the social risks to health, can best predict adverse health outcomes. Credit information bureaus turn to the diverse consumer and debt data that they own on most Americans to package and sell new predictive risk analytics directly to providers (Goodman 2016). Virtually any data can be used to make inferences on an individual's health status or predict risky health behaviors—a record of speeding tickets or car insurance claims can indicate dangerous driving, or frequent debit card swipes at a liquor store might indicate alcohol dependency. These data are integrated into credit risk scores, such as the FICO score, through a number of statistical methods. Some of the techniques used to build the FICO score have their origin in the lives of plants. Lyn Thomas has noted that the idea of the FICO risk score was born in the 1930s, and the algorithmic methods to build the score borrowed from genetic analyses that used statistical models to differentiate between iris varieties. A banking underwriter recognized that the same technique could be applied to distinguish good loans from bad ones (Thomas 2000, 151).

While these predictive instruments may use algorithmic techniques and statistical analyses similar to those used in health informatics research, much of the data used in health risk scoring are derived directly from clinical or biological information from millions of patients. For the predictive risk scores that the consumer credit industry uses, the industry makes inferences about individuals' riskiness based on a combination of these personal, alternative, and aggregate data.

As credit risk data platforms increasingly market their predictive products in healthcare, credit risk scoring may also assess patients' financial risks. For instance, the Fair Isaac Corporation (FICO), the company that owns the FICO score, has leveraged the data it already holds on millions of consumers to market a suite of predictive scores for healthcare use.[10] In addition to providing credit background on consumers, the largest credit bureaus—Experian, TransUnion, and Equifax—offer to healthcare providers both scoring products, such as the FICO Propensity Score for Healthcare, and sophisticated platforms, such as Experian Health, the information broker's "revenue cycle" platform for healthcare. A revenue cycle platform is essentially a patient debt collections system that uses health data against underinsured or uninsured patients. Healthcare providers can, in turn, run instant credit score checks on patients at the point of admission. These opaque credit data analyses are also increasingly marketed to healthcare providers for noncredit decision-making uses in medicine: a patient's credit score can be used to predict the likelihood of readmission or medication adherence. Equifax developed and marketed the use of credit scores to predict how individual patients may adhere to medical directives. In one Equifax white paper, researchers argued that while patient debt is often incurred beyond patients' control, healthcare providers, in their clinical decision-making, could nonetheless utilize the predictive power locked inside consumer data and a patient's debt:

> Since the mid-80s [lenders] have placed more confidence in the use of credit scoring to understand the client's behavioral tendencies. The predictive power that lies within credit information is so strong that most lenders use it as the main consideration when extending credit. Credit scoring has gained huge acceptance in this process because of its non-biased nature and the speed in which a decision can be made. Due to the increasing self-pay population, the healthcare industry can benefit from the lessons learned in the financial arena by placing confidence in the predictive power of consumer data (Equifax Predictive Sciences 2005).

These scores utilize marketing and consumer data, such as whether the patient owns a car, all considered "alternative" data, to predict if the patient will adhere to taking prescribed medications or will follow medical directives from their doctor. Increasingly, as well, healthcare providers are employing "wallet biopsies" on patients, enabled by the instant credit check platforms provided by Experian or Equifax, to establish a patient's ability to pay for certain medical procedures or determine what debt should be sent to collections.[11]

The credit bureaus, along with the creditors themselves, rely on predictive analytics, alternative data, and machine-learning techniques to assess risk (Hurley and Adebayo 2016, 184). Once collected, the data are analyzed to determine an individual applicant's level of riskiness or trustworthiness with an extension of credit. Yet as economic sociologist Martha Poon has argued, these analytical products, or "calculative objects" (2009, 657) as she calls them, are not constructed from a perfect set of data on a potential customer. Instead, they are

> parasitic and pragmatic constructions that make the most of information that is readily available at the bureaus as a resource for manufacturing prepackaged analytic products. These black-boxed statistical figures are in large part "behavioural scores." They do not seek to qualify static qualities of the person so much as they constitute a temporally responsive picture of consumer risk that is useful for tracking a person's ongoing relationship to credit. (2009, 658)

These analytics and data become the various products marketed to decision makers; these products are based on the consumer preferences and nonfinancial behaviors that credit bureaus sell to lenders and creditors. Often this is called "creditworthiness by association" because the risk assessment of an individual seeking credit is not based on his or her history of paying bills, or how many credit cards the individual holds (Hurley and Adebayo 2016, 151). Rather, the credit scoring is based on the predictive analytical methods borrowed from the consumer targeted marketing industry, which uses sophisticated algorithms to segment consumers into "lifestyle" or behavioral categories. Some in the credit industry argue that they need more robust data to analyze nontraditional debtors or people who operate in a cash economy (the "unbanked"), for whom there remains a dearth of credit history data. The credit industry is moving toward the mass and rather indiscriminate collection of data on all credit customers regardless of status. Furthermore, as noncredit decision-making increasingly utilizes credit data and scoring, such as when a potential employer runs a credit check on a job applicant or when a hospital uses a patient's credit score to predict her propensity to adhere to medical directives, there is a clear mission creep of credit, health, and marketing data with serious implications for personal privacy. This use of data—far beyond the subject's consent or knowledge—claims without rigorous evidence that these types of alternative data determine one's quality of life or future chances (Equifax Predictive Sciences 2005).

Data-driven decision-making based on predictive behavioral analytics, especially when those analytics are based on the behavior of millions of others, can produce negative outcomes for individuals who are subject to those predictive scores. For example, credit card holder Kevin Johnson was shocked to learn that consumer behaviors of complete strangers had brought down his credit score. Johnson, a Black business owner who took pride in his pristine credit history, was baffled as to why his card issuer, American Express, had reduced his credit limit by 65 percent. His credit limit was lowered not because he had missed a payment or defaulted on debt, but because creditors like American Express determined, based on behavioral analytics, that customers with bad credit also patronized the retail establishments he frequented (Hurley and Adebayo 2016, 150–51). Essentially, one's credit score may reflect not only a faultless personal history of paying bills on time, but also the perceived credit riskiness of relatives, friends, neighbors, or even complete strangers with similar consumer behaviors. In practice, an individual's credit risk is forecast based on otherwise personal (and at one time presumed private) aspects of his or her daily life: browsing history, Facebook friendships, stores frequented, and food purchased. For example, an individual who uses a credit card to purchase alcohol at a liquor store or makes "unhealthy" choices at a grocery store unknowingly generates data that credit assessors view as risky.

Increasingly, financial technology ("fintech") companies and startups offer analytical products that use these types of alternative data to create credit scores. Singapore-based startup Lenddo, for instance, collects information on whether a user's phone battery needs recharging, and it uses this data as an indicator of a person's level of responsibility and trustworthiness. The company then uses this information, combined with other alternative data they collect, to feed the scoring algorithm it sells to creditors hoping to reach the unbanked (Wei et al. 2016). As legal scholar Frank Pasquale noted in his discussion of China's Social Credit System, the US government may soon follow in scoring citizens based on consumer and other data, such as what they post on social media. During its infancy in the nineteenth century, the American credit reporting industry used such "alternative data"—gossip and hearsay collected from the friends and neighbors of those seeking loans or credit—to enhance the files they kept on debtors (Lauer 2017).[12]

Because the data, algorithms, software, and AI-driven technologies that support these decisions about credit users are proprietary, these scoring instruments remain unknown to those who use these scores to make decisions, and they are completely mysterious to those whose credit scores are subject to their influence. Since the analyses are based more on inferences made about an individual's consumer preferences and behaviors, as well as the preferences and behaviors of millions of others, and less on that person's credit history, it is much more difficult for consumers to know how what they may do or buy affects their credit scores. It is no longer enough for one to pay bills on time or be responsible with credit. Many in the credit industry place faith in the "unbiased" nature of these scores, arguing that algorithms can eliminate human error or prevent the replication of implicit racial or gendered biases that an individual underwriter might hold. "Objective" scoring is considered a more accurate reflection of credit risk, but this misplaced faith in automated decision-making may lead to dangerous consequences. As legal scholars Mikella Hurley and Julius Adebayo noted, "alternative credit scoring may ultimately benefit some consumers, but it may also serve to obscure discriminatory, subjective, and even predatory lending policies behind a single 'objective' score" (Hurley and Adebayo, 2016, 202–3). Recognizing that there is little fairness or transparency in how AI-generated credit scores are derived, Congress proposed legislation in late 2020 that would amend the 1972 Fair Credit Reporting Act to include provisions to make AI-driven credit scoring easier to understand for consumers—essentially, to break open the black box of the automated credit score.[13]

These alternative models are only as good as their data and their variables; if either are biased, the output will be biased as well (Browne 2015; Cheney-Lippold 2011; Dressel and Farid 2018; Kim 2017). When applied in the healthcare context, overreliance on decision-making based on credit or consumer data intensifies already problematic health inequities, such as denial of health services or inappropriate treatment based on a patient's financial status. Despite the blackboxing of the statistical methods and data that go into constructing credit scores, some in public health analytics suggest that the FICO score could say more about health outcomes than zip or genetic codes (Knapp and Dean 2018). Yet, the algorithm itself could exacerbate and replicate the systemic biases in banking and healthcare (Delgado et al. 2014; Mabry 2014).[14]

Take, for example, a widely used predictive algorithm that is fed by alternative data, specifically insurance claims data, rather than clinical data, to predict a patient's future healthcare needs. A risk predictive algorithm that the company Optum developed helps hospitals to identify high-risk patients with chronic conditions, such as diabetes or cardiovascular disease, so that providers can anticipate additional medical resources and care for those patients. Optum, a subsidiary of United Healthcare (UHC), bases its analytics on data from about fifty million insurance subscribers, including Medicare enrollees. The Centers for Medicare and Medicaid Services (CMS) subcontracts to private-sector healthcare insurers—including the Wall Street-traded UHC—to manage the medical care of Medicare and Medicaid subscribers.[15] Optum accounted for 44 percent of United Healthcare's profits, and Optum's global revenues exceeded one hundred billion dollars in 2019. Optum's value is not only in its possession of patient data, but also in what it does with that data: anything from securing and de-identifying or standardizing data, to using patient data in machine-learning or proprietary analytics. This range is what makes data monopolies, like Optum, so profitable.

Because Optum's data and predictive analytic platforms are proprietary, it is close to impossible to know how biases in either the data or the algorithms reproduce, and in some cases, deepen, systemic health inequities. One of the company's data products, a risk score physicians widely use to determine the level of care a patient should receive, uses United Healthcare claims data. When emergency medicine researchers tested Optum's algorithm, they exposed a significant racial bias in Optum's risk prediction product (Obermeyer et al. 2019). The authors found that Optum's predictive risk platform systematically discriminated against Black patients in predicting the medical care that a patient might require. The bias derives precisely from the tool's reliance on claims data—information contained in patients' medical bills rather than clinical records—to predict a patient's projected healthcare needs. Essentially the tool calculated how much a patient might *cost* the medical system rather than projecting a patient's medical care *needs* based on personal clinical data. And because so many clinicians relied on Optum's algorithm to make clinical decisions for patient care, the tool further marginalized vulnerable patients. These patients in many ways are the "missing patients" that Theresa worried about.

The predictive tool reflects the stark fact that the American health-care system spends less on Black patients, while evidence shows that these patients have comparatively a higher chronic disease burden, such as hypertension and diabetes (Crook and Peters 2008; Obermeyer et al. 2019). But this racist risk tool is part of a longer history of bias in health data in the United States. This history includes the insurance policies that covered enslaved people in the eighteenth and nineteenth centuries, as well as the fatalizing of Black life in the death tables actuaries use to calculate the riskiness of being Black, and thus, deny coverage in the twentieth and twenty-first centuries. In his research more than a hundred years ago on the health of Black Americans in Philadelphia, W. E. B. Du Bois noted America's "peculiar indifference" to the human suffering of Black citizens; it's not surprising to find that insurance data reflects the ugly reality that certain lives are valued over others (Bouk 2015; Du Bois 2007; Murphy 2013).

In Daniel Bouk's book *How Our Days Became Numbered,* an intriguing account of the early years of the life insurance industry in the United States, the historian noted that Americans first took up the culture of self-quantification and personalized risk management when they purchased life insurance policies in the early days of the twentieth century (Bouk 2015; Latour 1987). He called this quantification process "statistical individualism," and notes that it is performed alike by insurers and the insured (who share information about their health with the centers of capitalist calculation) (Bouk 2015). It is this particular type of individualism, or data "dividualism" in the Deleuzian sense, that feeds the scores that we increasingly live by and that we measure ourselves against. These numbers all rely on our digitized personal data. At times we part with our data willingly, say, when we share Fitbit data with a doctor. Mostly, however, we have no control and no idea how our data are built into biodata-based assets and used to score us, as is the case with instruments like the FICO score (Birch 2017; Ebeling 2016; Grundy et al. 2019). The current era of the quantified self—where we measure our bodies and health through wearable digital devices and platforms to develop scores for ourselves—has flourished in the fertile cultural ground of a widespread imperative to link our bodies to calculable risks, and to understand ourselves as risk centers for capital (Ruckenstein and Schüll 2017; Williams, Coveney, and Meadows 2015).

In a sense, we are doubly risky subjects of scores. The first way is in how the scores are built: a score puts us at risk of being "tossed out," of making us "missing at random." The second way is in how we become calculative subjects of scores—we are at risk of the score flattening or stigmatizing us. In the absence of regulatory oversight, the credit bureaus (soon to be joined by healthcare practitioners) seemingly require the surreptitious collection and combination of all these data crumbs and data exhaust so that we can remain "visible" and "legitimate" citizens within this intertwined regime of debt and medicine. In the following chapter, I investigate further how our data about our debts renders us visible.

6 Data Visibilities

> You're invisible, you're not there, they don't see you. I have my ID, and my name, and everything, but for them [I do] not exist on their screen because I don't have [a] credit history.
>
> —Akbar Amin, interview with author

"When I see my low FICO score, when I look at my credit card balance, I see love." A few years ago, a friend explained to me what her debt meant to her. Far from feeling shame or guilt, my friend expressed a sort of pride in knowing that by going into debt, she was able to care for and demonstrate her love for those closest in her life. A woman in her midthirties, Diana had worked all of her professional life in the service industries, first as a barista for Starbucks in her early twenties, and then, after she trained professionally, as a hairstylist for more than fifteen years.[1] At the time that she told me about what her debt meant to her, she was struggling to pay down more than twenty thousand dollars of credit card debt with the tips and small salary she earned as a contract stylist at one of Philadelphia's high-end salons. In the years leading up to the twenty thousand dollar balance, her debt went for small things, such as a vintage jacket she found for a friend, as well as to significant, important things, like to support a beloved relative as they recovered from an addiction disorder. But a large portion of that "debt love" went to save the life of Diana's cat. She used her credit card for all of the emergency veterinary visits as her pet became sicker, and then at the animal hospital, as she didn't have the seven hundred dollars in cash for his care. How Diana felt about her debt, how she saw that her life and her love were reflected in her debt, differed greatly from how what she owed on her credit card might appear to a mortgage broker or prospective employer when she applied for loan or a job. To anyone looking at Diana's credit scores or history of debt, she might appear like a risk, and perhaps not trustworthy enough for credit.

Debt is perhaps the aspect of modern American life that is most difficult to hide from the voracious collectors, brokers, and assemblers of data. And it is through the information collected about the debt that we carry that we tend to be defined and made visible. When combined with powerful digital tools, such as databases, targeting platforms, and predictive algorithms, this debt information is mixed together, dialectically, with the information of millions of others also in debt, and we are further defined, and made visible, by our data and the data of unknown others. Our data lives in credit scores, and have extensive "afterlives" beyond evaluating one's creditworthiness or financial stability. As discussed previously, in addition to a history of debts and whether they were paid on time, much of the data that comprises your credit score may be only tangentially about you, and may rely on information about complete strangers who shop at the same stores you do or happen to live in your zip code. In this sense, your credit score is a "dividual" amalgam (Cheney-Lippold 2017; Hayles 1999). As the intellectual property of credit-reporting agencies, the algorithms and data that comprise the credit score are secret, despite the Fair Credit Reporting Act demands for reporting transparency, so the information on what data are used, and how, exactly, these data are weighted and balanced is only as transparent as the law will allow. By packaging and repurposing our financial, consumer, and, increasingly, health information into scores or indicators of what we are likely to do or become, the credit bureaus construct us as predictive subjects of risk out of the data connected to our lives. These data are at the same time the assets, the products, and the raw materials of the consumer credit information industries, and as such, all of us, as consumers, debtors and patients, become their assets and products as well. How would you know what goes into the algorithm that determines your creditworthiness? This information is proprietary, a secret formula.

VISUALIZING DEBT THROUGH DATA

Increasingly Americans are governed, politically controlled, and subjugated by debt of all kinds. Generally, the personal debt that Americans owe is not due to extravagant lifestyles or living beyond one's means, but rather because so many are denied a means to live without going into debt to cover their basic needs. Indebtedness is a structural problem that negatively impacts individual households, so that it can feel like a personal failing rather than a social failure. Through a combination of financial innovations and shrinking economic opportunities, more Americans are in debt than ever before. Social policies established in the mid-1930s, such as the New Deal and the Federal Housing Administration, laid the groundwork for the popular use of debt to leverage the American Dream and middle-class aspirations (Bouk 2015). And since the mid-twentieth century, innovations in consumer credit and financial instruments—ranging from the invention of consumer credit cards in the 1950s, to the financialization of medical debt, and mortgage-backed derivatives in the early 2000s—that extended consumer credit widely, helped to fuel the mass indebtedness that many Americans live with (Debt Collective and Taylor 2020; Hyman 2012). The popularization of personal credit also went hand-in-hand with the neoliberalization of the American economy, and with the evisceration of the welfare state, the social safety nets built in the postwar era have evaporated since the 1980s, which has forced American households to rely upon expensive credit to pull them through hard times. Over the last forty years or so, for instance, the gap between workers' productivity for a company and what they are compensated has widened, while workers' wages have stagnated since 1972, and their incomes fail to keep pace with the inflation rate (Economic Policy Institute 2019). The combined US household consumer debt—including mortgages, home equity loans, auto and other personal loans, retail credit, credit card, and student loan debt—totaled close to fifteen trillion dollars by the end of 2020 (Federal Reserve Bank of New York 2021). Some of the largest personal debt loads for many

Americans are due to either student loans or medical costs, which can be incurred whether or not one has health insurance. By the beginning of 2021, close to forty-three million Americans were saddled with more than $1.6 trillion in educational debt, the only type of personal debt that cannot be discharged through bankruptcy. Medical debt accounts for two-thirds of personal bankruptcies in the United States, regardless of one's health insurance coverage (Himmelstein et al. 2019). Because medical debt can hide on credit cards and other forms of credit, like medical financing and payment plans, debt related to medical expenses can be disbursed and charged at different interest rates. Or it can be in collections. Debt has constrained, forestalled, and stigmatized the lives of many Americans. As such, social theorists and cultural historians have dubbed this distinctly American cultural phenomenon as a "debt society" and the United States as a "debtor nation" (Hyman 2011; Lazzarato 2015, 61).

At the root of the debt society is the rationale that somehow the personal debt that we owe is a judgment of our character, something that reveals a deep character flaw having to do with someone's greed, or laziness, or an addiction, or some other moral failing (Graeber 2014). In a debt society that is also a data-based one, how one's debt appears in personal data becomes all the more stigmatizing. Money in the form of currency is the ultimate debt instrument, as its value is in its promise that it will pay for a debt. It has no intrinsic worth. Money was invented to solve a societal problem with individuals' memory and trust as a way of recording and materializing debts and obligations. Currencies—everything from cowrie shells to bitcoins—can trace their origins to the recording systems of Mesopotamia. The priests at temples in ancient Babylonia kept accounts of transactions by printing on clay tablets or by notching sticks, that were then split, with one piece given to the debtor, and the other retained by the accountant or the creditor. These early ledgers were created as a recording device for transactions, obligations, and, ultimately, debt: what is owed and when is it to be paid back? They were one way to visualize the obligations of those in debt. Currencies developed out of these ledgers, and the value of currency, of money, is in the fact that it is a form of materialized debt.

In the modern era, debt is also materialized virtually in the form of digitized information, recorded through all of the data that we produce and that are collected within the debt-based economy. And through that data we are visible in the debt society. In this consumer, debt-based economy we are rendered legible through our data, and our lives are reduced to a series of scores that often precede us in virtually all aspects of our day-to-day existence. These numbers reflect more than creditworthiness; they also, often unfairly and without recourse, determine access to housing, healthcare, employment, education—in short, to all the things necessary to live a safe and fulfilling life, or even to just survive (Bouk 2017).

BEING SEEN IN DATA

How do we appear in data and who sees us? This depends. In his book *The Data Gaze* (2019), sociologist David Beer adapts Foucault's notion of the medical gaze developed in *The Birth of the Clinic*, where through technologies and techniques developed in Western medicine, the eye of clinical expertise objectifies the patient's body, abstracting it from the person's individuality and personhood (Foucault 2003). Beer argues that the data gaze "adapts and mutates when deployed in the codified clinic of data analytics"; while it "possesses a different kind of motivation, impulse, urgency and agenda [compared] to the 'clinical gaze.' . . . a key aspect of the data gaze remains the idea that it can reveal hidden truths that are otherwise invisible" (Beer 2019, 9).

The data gaze is also dependent on who deploys the gaze—who interprets the data—and through which infrastructure. If a pregnant woman in her thirties wants to buy a house, she will be visible through her data in different ways. A commercial data broker selling financial data to a loan officer in a bank might see a middle-class woman who will soon go through a major (and expensive) life event, the birth of a child, and her data may be made to conform to a marketing logic. But to a nurse working in emergency medicine and viewing the same person's electronic health record, she might be prediabetic, asthmatic, and at a high risk for pre-eclampsia. For a hospital using readmissions data, she might look like a risk to the hospital's bottom line because their Medicare funding metrics will be lower if she returns to the hospital for the same medical issue within thirty days of being discharged. To a data scientist working in healthcare, and who uses FICO scores as a variable for social risks in a predictive model like the ones discussed in chapter 5, she might appear as a complete abstraction, not definable as an individual but merely as a set of numerical values. Anmei, the CMU informatician we met in previous chapters, has exclaimed in moments of exasperated frustration, "Data scientists see numbers, we don't see people!" While healthcare providers do tend to see at least the part of the patient that is clinically represented in the medical chart, health informatics data scientists are not interested in the individual at all. Data scientists and analysts see tens of thousands of individuals as clusters of factors, values, and outputs.

Sociologist Janet Vertesi experimented with the question of how to maintain control over how we are seen, or how to remain invisible in data, when she kept her pregnancy off the digital grid and out of sight of the data-based economy. From the moment that she found out that she was pregnant, Vertesi decided not to let her unborn child become a part of the data amalgam, so she resisted announcing or sharing her news through social media, emails, or texts. She bought everything she needed for nine months at brick-and-mortar stores with cash, and used "burner" phones bought with gift cards that she also purchased with cash—all in an attempt to produce the fewest data breadcrumbs for brokers to use to make her legible to the data economy. By the end of her experiment, the credit bureaus and data marketers determined that her lack of data—her "thin file"—made her look like a criminal.[2] Vertesi's lack of visibility in her data rendered her suspicious and risky.

"A VERY AMERICAN THING": FOUR STORIES OF "DATA INVISIBILITY"

In a data economy that is based, at least in part, on information about personal debt, aspects of ordinary life that often remain invisible or forgotten are captured and materialized as legible data that define us as numbers and as scores. This reduction may haunt our entire lives. The cultural determinism of the debt society, and data's role in reinforcing that determinism, can be particularly apparent for newcomers to the United States, or for Americans who move away from the continental United States. The stories of the following four people, who struggled to make data about their debt visible (or invisible), illustrate how difficult it is to control even one's own data.

No Debt Makes You Appear "Untrustworthy"

Akbar moved to the United States nearly a decade ago from Jordan and recently received his US citizenship. When I first got to know him, he was looking for work and hopeful that his job search was nearly over. After what Akbar thought were successful in-person interviews with a car rental franchise, the manager he had met with initially offered him a job. But a few days later, after the employer had processed his application, which included details about his bank accounts, name, address, and driver's license, the manager called to say they had to withdraw the initial job offer. "Did they run a credit check on you?" I asked. He admitted, "I'm not sure." He figured they must have because of all the information requested and the forms he completed after the interview. Akbar couldn't remember the details of all the paperwork that he had signed, and since he really wanted the job, he didn't question whether he was required to sign all the documents. He had not heard of the Fair Credit Reporting Act (FCRA), so he would not have known that he had a right to refuse to have an employer run a credit check. At that point he couldn't imagine that a credit check would be a part of the job application.

The process of conducting a credit check, alongside a background check, has increasingly become a standard part of the job application process, and it is increasingly common even for low-wage positions (Traub 2014). Demos, a nonpartisan public policy organization, conducted research on the impacts of credit checks on low-income or chronically unemployed job applicants. One in four of their survey respondents reported that they had been required to complete a credit check as part of a job application, and one in seven had been denied a job because of the results of a credit check (2014, 3). These background and credit checks on new applicants comprise part of the larger "decisioning" products offered by information companies and credit bureaus. Decisioning is a branding device to promote a data platform that can be used, instead of evaluation by a human being (which is subject to biases, without question), to help make a decision about whether or not someone should be given a loan or credit card. Or, in Akbar's case, a job.

After Akbar relayed to me his initial confusion about why he might have lost the job he thought was a sure thing, I suggested that he didn't have a high "FICO score" or possibly no FICO score at all. He probably had a "thin" credit file for several reasons: he had been in the United States for a relatively short time, and his own culturally based preference was to save up for a purchase, as he had when he purchased his home, rather than to go into debt for it. In the algorithmic eyes of the debt society, no credit is bad credit.

A year later, Akbar and I met up again. After his job offer was rescinded, Akbar had asked other recent immigrants he knew if they had run into similar problems with credit checks and job offers. Many had, and they had advised him to apply for a credit card to "build up a credit history." Being in debt in order to have a credit history is a "very American thing," at least for Akbar, and something that he could not have conceived of, until he moved to the United States. After he applied for a credit card, however, he said he was offered a job—of course, only after the employer had run a credit check on him during the application process. Akbar noted that many of his immigrant friends had had similar experiences once they obtained credit cards and became visible to the debt data industry.

Invisible to the Debt Industry

Another recent immigrant to the United States, Austeja, noted that her perspective as a foreigner enabled her to see the cultural determinism in having to develop a credit history, being scored, and participating in a consumer culture. This societal pressure to submit oneself to being a subject of debt struck Austeja as very American. She was determined to resist it for as long as possible. Austeja had first come to the United States for graduate school at a land-grant school in the Midwest, and for the entirety of her graduate education, she managed to live without a credit card and without debt. While her income as a graduate student was very low, she kept her expenses low as well and was determined to resist the pressure to use credit of any kind. As a foreign student, she could not apply for student loans. She recalled feeling some relief about not going into debt like her American friends, but at the same time, she also felt financially insecure. In the event that something went wrong, like if she lost her teaching assistantship or something happened with her visa, she would have no safety net to fall back on.

Despite her conflicted feelings, for close to a decade Austeja was able to resist the debt industry. After she secured her present teaching position at a university on the East Coast, she reluctantly decided to apply for a credit card, in large part because having a credit history, and thus a credit score, would make it easier for her to rent an apartment or rent a car. When she accepted her teaching job, she felt that it was time to "grow up" and have contracts and debt in her own name. She considered the decision to be made visible by her credit history as less about her experience of having been invisible to creditors and felt financial insecurity, and more about an effort to be brought in from the margins, to be "legitimate."

Austeja noted the asymmetry of transparency in the relationship: she had to share her identity documentation, tax identification numbers, green card information, and anything about accounts with she might have with retailers or banks, all without hesitation, with this unknown and unknowable Big Other, a credit bureau (Zuboff 2015, 2019). As economic sociologists Marion Fourcade and Kieran Healy (2016) argue in their article "Seeing Like a Market," this lopsided relationship, where one party is expected to reveal all their private, intimate information to another party without a reciprocal expectation of exposure, is not necessarily a new phenomenon arising from the contemporary databased society. Fourcade and Healy note that its roots go back to the nineteenth century, when "American rating agencies developed methods to identify good credit prospects. They collected bits of information about the economic reliability of individuals and corporations" (10). Their intentions were not so different from today's data brokers: "capitalist markets and bureaucratic organizations shared an affinity for the systematic application of rules and measures that make the world legible so it can be acted upon" (10).

Living off the Debt Data Grid

Dan is another "data invisible" who managed to cultivate his invisibility for about four decades. Having grown up in the debt society as an American, and after dropping out of college in his early twenties, he decided that he would stay "off the grid" as much as possible. For a while, Dan was able to do this in Chicago. He took jobs that paid him under the table, bartered for services and housing when he needed to, paid cash for everything else, and at times (admittedly when he was much younger) scavenged for food and clothing. By his late thirties, about fifteen years ago, Dan decided to move to San Juan, Puerto Rico, a city that is now his permanent home. By only earning enough to survive, he remained out of the databases of debt. He avoided opening a bank account and for decades he never submitted a tax return. When his annual Social Security retirement estimate of exactly zero arrived in the mail, Dan confirmed that as far as the debt society was concerned, "they know I exist, but they think I am dead broke." Aside from having a Social Security number, which made him identifiable to the government, he was virtually undefinable to banking and consumer finance, as he had no financial or transactional data to feed the databases.

As Dan approached his midfifties, being off the grid was no longer tenable and he began to worry about the future for the first time in his life. He realized that he needed to save money for his children's college tuition, so he entered a teacher training program with the goal of finding stable, paid work as a teacher. And after surviving a series of devastating hurricane seasons, he felt that there was much more uncertainty, not only for the long term, but also in his day-to-day existence. In order to deal with some of that precarity, he planned to become visible through debt. He opened a credit card account with a very low credit limit and a high interest rate. Having never submitted a tax return, Dan figured he needed to do so before applying for jobs. He guesstimated his annual salary at around twenty thousand dollars and filed five years of back tax returns so that he would appear in more databases, especially when background and credit checks were performed as he went on the job market. His strategy worked, and a few months after submitting his tax returns as well as using the credit card for small purchases, he applied for and was offered a job teaching English literature at a local private school.

What's in a Name?

Martí never consciously chose to be invisible; it just happened to her through being misrecognized over the years. Martí grew up in the Río Piedras neighborhood of San Juan, not far from where Dan lives. There she spent her early life with her father, a well-known musician, and her mother, a literature professor who worked at the Río Piedras campus of the University of Puerto Rico, along with her two sisters. Martí now lives in Viejo San Juan, where she works as a librarian, but every weekend she and her partner, Pedro, go back to the old neighborhood to visit family and shop at the large farmer's market.

Martí's full given name is María Martita. Martí and her sisters share the same first name, María, a somewhat common naming convention in Puerto Rico, and since they were little girls, each sister has gone by her middle name. So, María Carmela goes by Carmela, and María Luisa is Luisa. Among family, friends, and colleagues, everyone knows Martí, but certain faceless entities, such as the credit industry, misrecognize Martí as one of her two sisters.

Martí first noticed her mistaken identity when she received a credit report she ordered from Experian. While her name and address were correct, Martí could not recognize any of the credit transactions listed. Her sister Carmela's mortgage, her sister Luisa's credit cards, all of the different addresses where each of her sisters had lived were all listed, but nothing of her own. This dilemma led her to muse, "But if my existence is not true, why don't [I] disappear? It's better to disappear. At least they can track you down. . . . I have thought a lot about that, of being invisible to the system."

Martí knew she had to correct this error, to make herself distinctly visible in the data and legible as an individual. For people in financially, emotionally, or physically abusive or exploitive relationships, or those trying to escape such conditions, the privacy exposures that Martí and her sisters faced could have been very dangerous (Littwin 2013). Fortunately for Martí, she has good relationships with both of her sisters, so she decided to enlist them to help her disentangle their data identities. When she spoke with each sister about how she was being misrecognized in data, neither seemed very concerned, so the task fell on Martí. Because her identity was mixed with those of her sisters, correcting it was not easy, but she managed to straighten out her credit history with Experian, which had sent the initial, mangled report. Before Martí got off the call with customer service, the representative told Martí she also would have to call TransUnion and Equifax and untangle her data from her sisters' with the other two credit bureaus. Martí gave up and decided to live with being semivisible.

In the end, Martí understood that her debt data identity was a fallacy, as she called it when we talked about her experience. But, she added, "you have to have [a data identity] in order to exist as an economic being." She pointed out that "the credit agencies that are supposedly... keeping safe that information, checking that it's correct" were not keeping up their end of the bargain: "it's not correct."

MAKING CREDIT INVISIBLES LEGIBLE THROUGH DATA

Akbar, Asteja, Dan, and Martí are what the credit reporting industry considers to be the under- or unbanked, or "credit invisibles"—at least they were until all four consciously made the effort to make themselves visible. These are people who, for a variety of reasons, for years remained "off the books" by not reporting their income, using cash instead of credit, and by refusing (or being unable) to engage with consumer banking or other traditional aspects of the economy. Credit invisibles are those people and households that do not have a bank account or use formal banking systems, such as credit unions, or personal banking instruments such as checks or credit or debit cards. The Consumer Financial Protection Bureau estimated that in 2015, twenty-six million Americans either had no credit history or had a record that was "unscorable" because the data were insufficient (Brevoort, Grimm, and Kambara 2015, 4).

"Unbanked" is an umbrella term to describe a diverse population of people who either underutilize banks or are off the banking grid entirely. It is expensive to be poor and unbanked. In order to be "banked," you need to have money, in the form of savings and cash flow, such as a regular paycheck or some other kind of compensation, like Social Security payments. Consumer bank accounts tend to have minimum balance requirements and charge fees when an account goes below that minimum balance, or if there are insufficient funds to cover a charge. There are fees associated with simply opening a savings or checking account and for credit or debit cards, there are fees associated with late payments or overdrafts. Consumer banks have cadres of data scientists who are tasked with innovating account fee structures in order to find new revenue streams by profiting off their customers. Because of the high costs associated with banking, unbanked people tend to participate in a cash economy, using pawn shops, payday loans, check cashing vendors, nonbank money orders, rent-to-own contracts, and informal social networks of family and friends to access cash and credit (Beard 2010).

The unbanked also tend to be individuals and families living at or below the poverty line, with high proportions of households headed by women, undereducated persons, people of color, immigrants, and young people. Banks and mortgage brokers have historically targeted particular groups, such as Black Americans or recent immigrants to the United States, for discrimination. Being unbanked for these groups has a long history, rooted in systemic racist policies such as redlining and housing discrimination.

For those who are invisible to the credit industry, it is extremely difficult—if not impossible—to apply for a mortgage to buy a home, a real asset often considered key to building generational wealth. Credit scores and credit background checks are also increasingly required for more than just applying for credit, as credit histories are required when applying for a job, renting an apartment, or even renting a car (Dean and Nicholas 2018). This debt visibility is almost a requirement for participation in any sector of American society.

If the consumer economy has a much more difficult time targeting and extracting value from data invisibles, we should recall that visibility and the generation of wealth have a fraught and violent history, especially in the United States. Many Black Americans experienced a forced visibility via the practice of redlining, which systematically denied them opportunities to build intergenerational wealth. Starting in the 1930s, government surveyors, along with the Home Owners Loan Corporation (HOLC) and local planners, colluded with banks to develop "residential security maps" that demarcated predominantly Black or immigrant neighborhoods in red ink as "no-go" areas for mortgage lending (Rice and Swesnik 2014, 936).[3] Though redlining was outlawed in 1968 through the Fair Housing Act, discriminatory housing practices, such as fraudulent rent-to-own contracts and homeowner's insurance or credit interest rates based on zip code discrimination, remain (Cohen-Cole 2010). To this day, in addition to ongoing discrimination in lending and housing, majority Black neighborhoods and primarily Latinx or immigrant communities that were historically redlined tend to experience higher poverty rates, higher instances of cancer and other health complications, and lower chances for families to build intergenerational wealth (Bartlett et al. 2018; Beyer et al. 2016; Jacoby et al. 2018). Forced visibilities in the context of Puerto Rico, where Puerto Ricans themselves are subject to what Ren Ellis Neyra calls US sovereignty's trespass, must also consider that it is not just individual debt or liabilities that are made visible, but colonial debt and ongoing dispossession on the archipelago as well (Ellis Neyra 2020, 58). The afterlives of discriminatory practices such as redlining or US economic imperialism are reborn in data-based decisioning products, like the credit score.

In Edward Tufte's book *Beautiful Evidence*, the information designer argues that simply making something visible is not enough; there is a moral obligation to use the visual to explain, to educate, and to provide evidence (Tufte 2006). In one compelling example, he displays the now famous slave ship schema, which depicted how captured Africans were crowded below deck. The Religious Society of Friends (Quakers) in London used this image to make a visual argument against slavery, exposing the cruelty and inhumanity of the transatlantic slave trade, and to demand the abolition of the trade and enslavement of Africans. While the abolitionists' visual argument was an appeal to those in power to recognize the inhumanity of the system they presided over, it was also a moral argument that conveyed the trauma and lives lost to the brutality of a global economy based on making human beings into commodities. There is a similar flattening of the fullness of one's life into quantifiable bits of information in order to be visible in data, and a comparable moral argument when one sees the visual evidence of practices such as redlining.

An argument often made in the promotional literature of the credit and banking industries is that in order to build wealth, it is beneficial to be visible in the debt society. Unlike Tufte, who proposes that visibility should be used to make a moral argument or to provide evidence of injustice, the industries that make commodities of our data make no such claims. The way that one is seen by their debt is through their credit information data. Banks and associated financial services firms such as credit bureaus see credit invisibles and the unbanked as yet another untapped market, and this industry has developed new data-based products in order to make them visible. The key to making the unbanked visible and legible is not just any data, but alternative data: those bits of information scraped from our daily interactions and activities and their afterlives in the algorithms of data brokers.

One firm that claims to make the unbanked visible to lenders is VisualDNA, a small marketing information startup absorbed by the Nielsen Corporation (best known for its television ratings). VisualDNA's core business is collecting data through OCEAN personality profiling to provide psychographic and other data for credit risk scoring of those historically locked out of banking: it calls this model "physical people as digital data." On one webpage used to market its services to lenders looking to acquire new customers, for instance, VisualDNA discusses the science behind their data products: "Using the latest psychological research, VisualDNA has developed a methodology which accurately depicts a physical person as digital data by assessing personality, values and attitudes. That enables us to predict specific financial behaviors on the basis of scientifically valid measures and psychological constructs both based on solid models; such as the Big Five model of personality."[4] Through the use of "personality quizzes" that categorize credit invisibles into the OCEAN personality model (famously weaponized by Cambridge Analytica, as discussed in chapter 4), VisualDNA claims that its models make the unbanked more profitable and less of a risk to lenders. Such practices also force a peculiar visibility onto those who are seeking credit.

In a report to investors in spring 2019, Experian announced its success in collecting new data from more than six hundred thousand consumers formerly considered "credit invisibles" in just six months. The Experian Boost data platform promises those who are underbanked that it will "instantly raise" their credit score, in exchange for sharing bill-paying and other transactional data. The Boost website features stock photos of people of color, virtually all female and under forty years old, and the dashboard enrollment tool has the distinct look and feel of gamified social media. Once a consumer enrolls on the website, the enrollee gives Experian permission to access information from their banks and to collect transactional and payment data from their mobile phone carriers, cable television and streaming services, and other vendors that traditionally do not report on-time payment histories to the credit bureaus. Experian is after are what are considered "nontraditional" data and yet are increasingly being collected and deployed to render the "credit invisibles"—low-income or unbanked individuals—visible to the credit system. In one Boost advertisement, titled "Stampede," a white man coded as rural and working class through his denim work clothes and boots, looks at his old pickup truck and says rather forlornly, "I wish I could buy a new truck." After the actor uses the Boost app on his phone, the shabby pickup transforms into a shiny new truck, which he can metaphorically use to drive out of the desert of social determinism and poverty.[5] Under the guise of empowering consumers, especially those locked out of credit historically, credit bureaus encourage consumers to share more of their data, falling deeper into the data economy.

Since at least the 2008 economic collapse, in which millions of lower- to middle-income Americans lost their homes or saw their savings wiped out when the mortgage-backed derivatives bubble burst, the banking industry has been under pressure to redress discriminatory lending and consumer banking practices. This insistence on more social equity and fairness in banking has only accelerated since the pandemic and the Black Lives Matter uprisings in 2020, both of which underlined the historic, systemic inequity in generational wealth for Black and brown families as compared to white ones. Nancy Levine, a former banking regulator who now works on a consumer bank's fair lending team, explained to me, in an interview about data use and fair credit practices, that her group is turning to alternative data products, like Experian's Boost, to make nontraditional data perform two functions. The first is to find credit invisibles who could be considered creditworthy outside of the models that make up the FICO score. The second is to use these newly found creditworthy applicants as proof that her bank is not only compliant with, but innovating on, the fair credit principles. For Nancy, the use of alternative data to improve an individual's FICO score by a few points, or to make people visible to the debt-based economy, is a question of equity. For those groups who have been historically excluded from access to credit, such as people of color, women, and low-income people, accessing more credit may be a good thing. But if the bargain requires that more people take on debt, rather than build wealth—and all through giving up more of their data to these credit bureaus—maybe the bargain is not about justice but about control of one's visibility.

The demand to make oneself visible, or the decision to remain invisible, are expressions of power. Poet and philosopher Édouard Glissant argues in his essay "For Opacity" that power requires complete transparency of its subjects, and thus this demand is made always as a bid to "understand" the subjects. Yet the demand is necessarily reductive of the subject herself in order to make her an object of comparison or analysis (1997, 190). The challenge is not to be reduced to and flattened by data, and to resist being mixed into the digital amalgam (Glissant 1997, 192). Wendy Chun writes about the ideological shaping of software, especially the gender and race ideologies that reinforce the power behind computing—primarily late capitalism and neoliberalism—in her book *Programmed Visions: Software and Memory* (2011). Visibility in software interfaces, far from making the power transparent, helps to hide its control. Rather than let us see what is invisible, she argues that as computers shrink (from servers to laptops to smartphones and watches), they become more opaque, more like black boxes, and that "we the so-called users are offered more to see, more to read. As our machines disappear... the density and opacity of their computation increases" (Chun et al. 2011, 17).

Chun argues that "software, through programming languages that stem from a gendered system of command and control, creates an invisible system of visibility, a system of causal pleasure. This system renders our machine's normal processes demonic and makes our computer truly a medium: something in between, mystical, channeling, and not entirely trustworthy. It becomes a conduit that also amplifies and selects what is at once real and unreal, true and untrue, visible and invisible" (2011, 18). The opacity of these black boxes enables us to be read even as we read. Those black boxes also include the credit scores built with our purloined data.

In her book *Beautiful Data: A History of Vision and Reason since 1945* (2014), media historian Orit Halpern examines the concept of visualization as it is operationalized through data, which she contends has had a profound influence on how we record and perceive information. She notes that through the development of scientific instruments and technologies that render information by means of numeric measurements or characterization, the contemporary understanding of visualization makes "new relationships appear and produce new objects and spaces for action and speculation" (2014, 21).

But as social theorist Andrea Mubi Brighenti (2017) argues, the visible is not simply revealing "new relationships"—the relationships that are revealed are primarily to what remains invisible. Visibility is only possible as construction or a simulation of what is unseen. He makes a key distinction between what is visible and what is seen:

> Now, the visible is not the seen, just as the invisible is not the unseen.... For its part, the visible is beyond the seen and the unseen, beyond attention and interest; if ever, attentive phenomena are tropisms and vectors inside the visible.... The latter is, in fact, not a relation of opposition and mutual exclusion. On the contrary, the invisible constantly contributes to the visible, it adds to it. In this sense, the invisible is no less operative than the visible. The invisible is the visible without a theme. This is why... visibility itself can only be understood as a notion of *virtuality, not actuality.* (2017, 2; my emphasis)

Visibility in data could be understood as a virtual manifestation of life—not life itself, only a simulation of life. Data, once visualized and on display, become the new objects for action, outlining and defining not only new relationships but also an internally defined relationship *to itself.* Once data are made visible and legible, the relationship to actuality is virtual and open to speculation by all sorts of actors in the data economy. But what is revealed when something is made visible? A distortion of our lives, which are flattened to the surface of data; what is made visible, what is reflected back is only the brutality of a society controlled by debt.

In promoting its social media data analytics tool to advertisers, for instance, Experian describes the tool's ability to turn "the activity generated from over 90 million social media users into meaningful, *actionable* intelligence. Delivered through a web-based dashboard and customized to your audience, the Social Media Analysis helps advertisers and agencies easily *visualize* key findings."[6] Visualization is not simply a process of making an unseen relationship legible; visualization should be understood as an apparatus or "conditioning infrastructure for how subjects come to be known to power" in a Foucauldian sense (Halpern 2014, 24). That is, making something or someone "visible" always involves the less powerful being forced into visibility by the more powerful.

Once you are made visible to the digital systems by being made into data, there is a second process to make your data legible, your *life* legible, by making the data readable by computers and by their human interpreters. Algorithms force these disparate bits of information into a particular narrative or category (genomic data, financial data, behavioral data, or psychographic data) with a specific legibility that gives the data meaning for a user. Your data, once scraped, collected, or purloined without your consent or knowledge, doesn't just sit in an Excel spreadsheet file or a data warehouse. There, it is dead, and it is considered data refuse or data exhaust, the waste material of the data industry. In order to make data useful, it must be mobilized and operationalized, often in very different contexts from the ones in which it was collected. Once it is on the move, through its repackaging and disclosure to other third parties, it will be transformed to be made legible over and over again in a variety of platforms. Your data operationalized in an electronic health record will be "read" differently as compared to how a marketer might read your data. The data or credit invisibles are made visible and legible by forcing them to fit into this data-debt regime. You are made visible by your data and made legible by data analysts.

In one sense data are information derived from measurements or statistics that form the basis of reasoning or making a decision about something. In another sense, data are information that may be redundant and need to be processed in order for them to have meaning. Credit bureaus target and extract the data that make our lives legible, jerry-rigging our lives to conform to the logic of data, no matter how incongruous our lives may look in the form of a FICO score. Numbers or data themselves may be an "accurate" depiction of a transaction or of our lives, but what is important is that these numbers, these scores, give a certain legitimacy and visible demonstration of the social authority that the credit bureaus hold in the debt economy.

The Debt in America interactive online map, is a good example of how the social authority of the institutions that own our data can be challenged and even subverted. The data assets regarding American's personal debt obligations that are owned and controlled by credit bureaus can be analyzed to make us legible as risky subjects, but when used in another context, they have the potential to subvert the debt industry's control over how we are made legible in our data and who profits from that legibility (Hou et al. 2021). The Debt in America online tool, built by the nonpartisan think tank the Urban Institute, was constructed to visually demonstrate the ways that personal debt can "reinforce the wealth gap between white communities and communities of color" (Urban Institute n.d.). The Debt in America data map uses anonymized datasets of 98 percent of American households' and their debt information, purchased from credit bureaus, like Experian or Equifax. These datasets include information about all of the personal debt that most Americans hold, including medical and educational debt, personal loans, and balances on credit cards. Because some of the credit reporting agencies, like Experian, also provide revenue cycle products to the healthcare industry, the dataset also includes information about medical debt in collections—such as unpaid hospital bills that have been sold to third-party debt buyers for pennies on the dollar.[7] Close to forty-three million Americans have medical debt in collections, and 62 percent of personal bankruptcies are due to unpaid medical expenses (Consumer Finance Protection Bureau 2014, 2017; Himmelstein et al. 2009). Users of the Debt in America data map can toggle between overall debt, medical debt, student debt, and auto debt, and compare debt loads regionally. The map makes visible that most Southern states have much higher rates of personal debt compared to other states, and that white communities experience significantly lower debt loads compared to communities of color. The map also allows users to compare categories of the debt that are in collections, which helps to visually delineate the impacts of racism on personal debt and the wealth gaps between white households and households of color.

The Debt in America data map uses the same data that the debt industry uses, but by making them visible and in relation to one another in ways that are in opposition to the debt society, the Urban Institute makes a compelling visual and moral argument for social equity and reparative justice for the historic and systemic harms wrought by white supremacy and racism. In this sense, the Debt in America data map subverts the power of the data gaze and bends it towards justice for the subjects of data and debt.

The trick then, for we who are subjects of data and subjects of debt, is to remain opaque, that is, to remain complex and not to be flattened by the demands of the data-based debt economy. It is also to demand that data about us cannot be used without us; not only must we control how we are made visible and legible in the data-based economy, we should demand that our data visibilities are in the service of justice. It is not an easy maneuver to pull off anymore, as the demand to be transparent, to be visible through our data and reduced to a score, is a demand that increasingly is impossible to resist. The secret is to not fall into the amalgam. But it is virtually impossible in the databased debt society to withstand becoming a "dividual"; the best that one can do is to hope not to be plucked out of the aggregate by late capital's vultures and predators.

Epilogue

AFTERLIFE

By now, after more than a year of teaching, meeting, and attending seminars and conferences on Zoom, I have developed muscle memory. Log on, turn off my camera, stay muted, and take an occasional sip of coffee from my favorite mug. Switch on my camera and microphone only to speak or to demonstrate that yes, I am still in front of my computer. Because of the pandemic, I have not been back to the field site at Central Midwest University (CMU) in more than a year. But because of the pandemic and because of the marketplace dominance of meeting platforms like Zoom, I've still been able to virtually visit Theresa's lab meetings and participate in some of the symposia sponsored by the Consortium. Though I am in Philadelphia to attend today's meeting, most of the attendees are geographically closer to the Division, the Midwestern city street not far from Theresa's lab, where I first met Steve, Emma, and Carl, some of the founders of the Consortium. Many of the screens are also silent, black and emblazoned with the user's name, just like mine.

The past year's public health crisis, spurred by the pandemic, has tested the stability of the data infrastructure and demonstrated the crucial role that responsible data sharing plays in protecting and saving lives. Through this trying year, the Consortium has matured, and it is in the next phase of advancing its mission to build a cross-sector data infrastructure that centers personal data as a public good, and also to use data for social good. The Consortium is convening today's online meeting to introduce some of the organization's newest collaborators from the nonprofit and academic worlds, including Data.org, the Center for Inclusive Growth, and CMU's school of social work.

Steve notes that the Consortium was founded to bridge the data and expertise fragmentations that exist in social services by working "on a cross-sector strategy around this shared data infrastructure piece that can connect public health, social services, education and community development and trying to do so in a way that is both more efficient and more impactful, because we're doing it as one collective whole, rather than sort of piece by piece." After Steve opens the floor to the new partners to introduce themselves and their organizations, he reminds attendees why we are collected together today: "because no one believes that data for data's sake is sort of just a good in itself, but [we want] to understand how data can be leveraged... to impact [and] improve people's lives." He adds that the Consortium does "the work of data sharing and data analysis across organizations, but also [we] support strong data actors, just to do things like we're doing today, to make sure people have access to the capacity, the support, and the connections that they need to do data work effectively in their community."

One of those strong actors is undoubtedly the Center for Inclusive Growth, MasterCard's primary philanthropic foundation, which shares its oceans of financial, debt, and consumer data with the public sector to address social inequity through community investment and economic redevelopment in the United States and globally. Since the Consortium's official launch, MasterCard is the first multinational financial services corporation to partner with the nonprofit organization. The alliance with such a powerful corporate player that handles and holds debt data seems to be a turning point for an organization dedicated to cross-sector data sharing as a social good. Up until this point, the Consortium's collaborations were either with other nonprofits working in social services and academia, or agencies in the public sector, like regional public health departments, and public assistance or housing authorities.

With the Center for Inclusive Growth partnership, the Consortium can now leverage the personal financial and consumer data of millions of low-income residents of the region—data assets that are owned by the financial services behemoth—to advance part of its mission to use data as a public good and in the service of doing societal good. There is little doubt that the MasterCard data assets that the corporation shares with its philanthropic arm include transactional data, information about purchases of goods and services, from the MasterCard-branded debit and EBT cards used by enrollees and recipients of SNAP, unemployment benefits, and public assistance.

Steve clicks his mute button and hands over control of the screen to Casey, the senior vice president of the Center for Inclusive Growth. After giving some background on the organization's mission, primarily that the philanthropy promotes financial security and economic mobility through data-driven solutions, he shares a presentation to demonstrate how the Center for Inclusive Growth helps the public sector use MasterCard proprietary data for what he calls "social impact," through products like its "inclusive growth score" tool. He clicks on the data dashboard to show how the score can be used to enrich datasets and refine understandings about what social determinants are driving poverty in a given census tract. As he clicks on the dashboard's buttons and drop-down menus, Casey explains that the tool uses a mix of MasterCard data and data obtained from external sources. He chooses a census tract near CMU's main campus on the menu, and clicks another menu to choose inputs from the Center for Inclusive Growth's data assets to enhance the tool. With the inclusion of this alternative data, Casey explains that this additional information suggests that the poverty rate in the census tract seems to be associated with the percentage of females living above the poverty line and by the enrollment rate for early education.

As he wraps up his presentation on the proprietary "inclusive growth score," Casey gives a final pitch on what role the Center for Inclusive Growth, and ultimately, MasterCard, hopes to play in the public sector. He says that by leading the charge to provide thought leadership MasterCard can:

> challenge the status quo and evangelize for the underserved. And the clips you're seeing on screen are just a few of the ways in which were publicly evangelizing for change, and more broad-based adoption of data solutions in the social sector. And while we feel compelled to lead the charge and to use our own data to advance inclusive growth, it's critical that we begin to

> expand beyond our own data and research, and work with key partners to build the capacity of the social sector. . . . We want to make sure that we are working to help promote [the tool's] development that can be used by the social sector.

This partnership might be of some benefit to Consortium members because with the MasterCard datasets members can enhance their own data, and possibly provide richer and more nuanced insights to support decision-making for social impact. But the partnership still represents an inequitable bargain that is all too often asked of our society's most vulnerable: the indebted, the unbanked, the invisibles or the hypervisibles. Certainly, the data contained within MasterCard's datasets are the assetized traumas of the indebted millions, forced to draw upon ever-shrinking social benefits due to layoffs or catastrophic health emergencies, or simply because of the cruel optimism of neoliberal trickle-down economics; no matter how many bootstraps are pulled, one can never seem to get out of poverty (Berlant 2011). MasterCard extracts these data in the context of surveillance capital, a casino where the house always win. Rather than making the unbanked visible, these data serve as a barometer to measure the cruel inhumanity of neoliberalism, of the debt society's assault on the poor, dispossessed and the indebted. Corporations like MasterCard, Google, and Microsoft contribute to the conditions of poverty, and help to create the indebted class, then use those economic circumstances, which they helped to create, to build their charitable activities. Just like they were for Rockefeller or Carnegie, the robber barons of America's Gilded Age, these corporate philanthropic arms are tax havens for data monopolies; they also serve as a public relations performance, putting some shine on the business of profiting off dispossession and debt.[1] If data are to do any good here, it should be to towards a reparative justice to reverse the brutal inherences of the debt society.

In the moments after I logged off of the meeting where I learned about this new development between the Data Consortium and the debt society Goliath, MasterCard, I found my thoughts drifting back to Ana. It has been close to five years since Ana passed away, and as I reflected on the intervening years, it seems that while so much has changed, or at least deepened—America's divisive and deadly politics intensified under the

Trump Administration, the gaps of economic inequality continue to grow, and the structural injustices endure—but some things most certainly continue as when she was alive. I am fairly sure that Ana's data, and through her data, she herself, still endures somewhere in the debt society's databases, algorithms, scores, and statistical models. Ana's data survived her death and have been transfigured, and are certainly alive. Though I don't know exactly what happened to all the data collected on Ana as she went through the medical treatment ordeal in her efforts to survive brain cancer, I can make some educated guesses. All of her interactions with profit-driven medicine would have been captured in her electronic health record, recorded as diagnostic and procedural codes. These data would have been stored, amassed, and aggregated with those of millions of other patients in healthcare data oceans. They would have been shared with any insurers or payers financially responsible to cover her care, after accounting for her co-pays and any treatments or drugs not covered by insurance of course. Certainly, her cancer diagnosis was known to the data industry, to companies like Experian, and those in the credit industry, such as MasterCard, as they would have collected information on the way that she paid for her chemo and surgeries, even if she was insured. This data, combined with the information they already had on her toxic debt, like the student loan payments that went into arrears as she struggled to survive cancer, would have scored her as "risky." Those companies in the credit-reporting industry, and countless other third-party companies, even corporate philanthropic foundations such as MasterCard's Center for Inclusive Growth that are well outside of healthcare system, would have had access to her health data—data ostensibly locked under privacy protection and secured in her health records. And the data miners would have found the crowd-sourced fundraiser made by her friends, and whether she shared a link to it on her social media accounts. My hope is that Ana's medical data, at least, has possibly lived on in an algorithm to help save the lives of other patients diagnosed with brain cancer. I'm not sure if Ana was on Twitter or Instagram. I only know about her Facebook account, as we were "FB friends." We still are. Yes, the data-debt society would know where to find her, even in death.

After she died, many who loved Ana while she was alive posted messages of mourning, sadness, and disbelief that she was gone to her Facebook account, which was never deactivated after her death. Every year since, on the anniversary of her birth—because Facebook reminds us—her loved ones post their birthday wishes to her, recount stories and memories from her life, and tell her how much they miss her. In one recent post, a bereaved friend uploaded to her Facebook page a touching video of herself walking down Ana's old street, as a birthday gift. While everything that makes our lives—our health, relationships, feelings, experiences, memories, senses—may be fleeting, the afterlives of our data continue on in the databases, in the risk scores and predictive models of surveillance capitalism. Our data are eternal.

Notes

INTRODUCTION

1. Lizzie Presser, "When Medical Debt Collectors Decide Who Gets Arrested: Welcome to Coffeyville, Kansas," *ProPublica*, October 16, 2019, https://www.kcur.org/post/when-medical-debt-collectors-decide-who-gets-arrested-welcome-coffeyville-kansas. Wendi C. Thomas, "Low-Wage Workers Are Being Sued for Unpaid Medical Bills by a Nonprofit Christian Hospital That Employs Them," *ProPublica*, June 28, 2019, https://www.propublica.org/article/methodist-hospital-sues-low-wage-workers-medical-debt. Mary Katherine Wildeman, "A Loophole Lets SC Hospitals Take Millions from Residents' Tax Refunds for Unpaid Bills," *Post and Courier*, April 20, 2019, https://www.postandcourier.com/business/a-loophole-lets-sc-hospitals-take-millions-from-residents-tax/article_92a381a4-4b77-11e9-b439-ffe02586b0af.html.

2. Shefali Luthra, "When Credit Scores Become Casualties of Health Care," *Kaiser Health News* (blog), May 9, 2018, https://khn.org/news/when-credit-scores-become-casualties-of-health-care/.

3. Steve Benen, "GOP Senator: Health Care Coverage Is 'a Privilege,' Not a Right," *MSNBC*, October 2, 2017, http://www.msnbc.com/rachel-maddow-show/gop-senator-health-care-coverage-privilege-not-right. After I contributed to Ana's crowdsourced fundraiser, between 2017 and 2021 I contributed to my cousin's fundraiser for a kidney transplant, and gave money to another friend's GoFundMe to help pay for his $195,000 medical bills for pancreatic cancer. These

were in addition to contributions that I have made to strangers' medical fundraisers to pay for insulin, surgeries, and, in a few cases, funerals.

4. Karen Weise, "Amazon's Profit Soars 220 Percent as Pandemic Drives Shopping Online," *New York Times*, April 29, 2021, https://www.nytimes.com/2021/04/29/technology/amazons-profits-triple.html.

5. Charles Duhigg, "How Companies Learn Your Secrets," *New York Times*, February 19, 2012. Alexei Alexis, "Big Tech's Data Control Faces Antitrust Scrutiny at FTC," *Bloomberg Law*, February 27, 2019, https://news.bloomberglaw.com/mergers-and-antitrust/big-techs-data-control-faces-antitrust-scrutiny-at-ftc. Gerrit De Vynck, Cat Zakrzewski, Elizabeth Dwoskin, and Rachel Lerman, "Big Tech CEOs Face Lawmakers in House Hearing on Social Media's Role in Extremism, Misinformation," *Washington Post*, April 9, 2021, https://www.washingtonpost.com/technology/2021/03/25/facebook-google-twitter-house-hearing-live-updates/.

6. See, in particular, Title V of the GLBA, which includes the Privacy Rule, the Safeguard Rule, and the Pretexting Rule. Privacy of Consumer Financial Information Rule, Gramm-Leach-Bliley Act of 1999, 16 CFR 313. These rules are administered by the Federal Trade Commission; they first passed in May 2000, with the most recent amendment to the Privacy Rule in December 2021, effective January 2022.

CHAPTER 1. TRACING LIFE THROUGH DATA

1. Isobel Asher Hamilton, "'Our Product Is Used on Occasion to Kill People': Palantir's CEO Claims Its Tech Is Used to Target and Kill Terrorists," *Business Insider*, May 26, 2020, https://www.businessinsider.com/palantir-ceo-alex-karp-claims-the-companys-tech-is-used-to-target-and-kill-terrorists-2020-5.

2. Edward Ongweso, "Palantir's CEO Finally Admits to Helping ICE Deport Undocumented Immigrants," *Vice*, January 24, 2020, https://www.vice.com/en_us/article/pkeg99/palantirs-ceo-finally-admits-to-helping-ice-deport-undocumented-immigrants. Spencer Woodman, "Palantir Provides the Engine for Donald Trump's Deportation Machine," *The Intercept* (blog), March 2, 2017, https://theintercept.com/2017/03/02/palantir-provides-the-engine-for-donald-trumps-deportation-machine/.

3. Marisa Franco, "Palantir Filed to Go Public. The Firm's Unethical Technology Should Horrify Us," *The Guardian*, September 4, 2020, http://www.theguardian.com/commentisfree/2020/sep/04/palantir-ipo-ice-immigration-trump-administration.

4. "Preview: 'Axios on HBO' Interviews Palantir CEO Alex Karp," *Axios*, May 25, 2020, https://www.axios.com/axios-on-hbo-palantir-alex-karp-9788cf3e-6c0c-4794-bc05-c472b4bc98fa.html.

5. Mark Harris, "The Lie Generator: Inside the Black Mirror World of Poly-

graph Job Screenings," *Wired*, October 1, 2018, https://www.wired.com/story/inside-polygraph-job-screening-black-mirror/.

6. Ziad Obermeyer, interview by Ira Flatow, "Can an Algorithm Explain Your Knee Pain?" *Science Friday*, May 14, 2021, https://www.sciencefriday.com/segments/algorithm-healthcare-pain/.

7. Cade Metz and Daisuke Wakabayashi, "Google Researcher Says She Was Fired over Paper Highlighting Bias in A.I," *New York Times*, December 3, 2020, https://www.nytimes.com/2020/12/03/technology/google-researcher-timnit-gebru.html.

8. Russ Kick, "Exclusive: ORR's Spreadsheet on Pregnancy of Migrant Girls in Its Custody," *Alt Gov 2*, May 3, 2019, https://altgov2.org/orr-pregnancy-spreadsheet/.

9. However, laws concerning the age of consent vary among states in the United States, as well as among these states and the countries that the girls originated from.

10. Garance Burke, "Federal Agency Says It Lost Track of 1,488 Migrant Children," AP News, September 20, 2018, https://apnews.com/aad956b7281f4057aaac1ef4b5732f12.

11. "Judge Orders Trump Administration to Stop Blocking Abortion for Two Immigrant Women," American Civil Liberties Union, press release, December 19, 2017, https://www.aclu.org/press-releases/judge-orders-trump-administration-stop-blocking-abortion-two-immigrant-women.

12. Katyanna Quach, "MIT Apologizes, Permanently Pulls Offline Huge Dataset That Taught AI Systems to Use Racist, Misogynistic Slurs," *The Register*, July 1, 2020, https://www.theregister.com/2020/07/01/mit_dataset_removed/.

13. Kashmir Hill, "Wrongfully Accused by an Algorithm," *New York Times*, June 24, 2020, https://www.nytimes.com/2020/06/24/technology/facial-recognition-arrest.html.

14. Hill, "Wrongfully Accused."

15. Timnit Gebru was an early organizer and lead in the American Computing Machinery's Conference on Fairness, Accountability, and Transparency, whose inaugural conference was organized in 2018 and held in New York City.

16. Matthew Desmond, "American Capitalism Is Brutal. You Can Trace That to the Plantation," *New York Times Magazine*, 1619 Project, August 14, 2019, https://www.nytimes.com/interactive/2019/08/14/magazine/slavery-capitalism.html.

17. SNAP, formerly known as food stamps, is a US federal program, dispersed and managed at the state level.

CHAPTER 2. BUILDING TRUST WHERE DATA DIVIDES

1. While this is a pseudonym for the street, the corridor's actual colloquial metaphor of cultural, economic, and racial divides is widely known and used by residents.

2. All names and locations throughout the book have been anonymized with pseudonyms.

3. Infant and maternal mortality rates in the United States are the highest of all developed countries. Rabah Kamal, Julie Hudman, and Daniel McDermott, "What Do We Know about Infant Mortality in the U.S. and Comparable Countries?" *Peterson-Kaiser Health System Tracker* (blog), October 18, 2019, https://www.healthsystemtracker.org/chart-collection/infant-mortality-u-s-compare-countries/.

4. The creation of a national network of connected regional health information exchanges was one of many goals of the HITECH Act.

5. Cade Metz and Adam Satariano, "An Algorithm That Grants Freedom, or Takes It Away," *New York Times*, February 6, 2020, https://www.nytimes.com/2020/02/06/technology/predictive-algorithms-crime.html.

6. Drew Harwell and Eva Dou, "Huawei Tested AI Software That Could Recognize Uighur Minorities and Alert Police, Report Says," *Washington Post*, December 8, 2020, https://www.washingtonpost.com/technology/2020/12/08/huawei-tested-ai-software-that-could-recognize-uighur-minorities-alert-police-report-says/.

7. Xinyuan Wang, "China's Social Credit System: The Chinese Citizens' Perspective," *Anthropology of Smartphones and Smart Aging (ASSA)* (blog), December 9, 2019, University College London, https://blogs.ucl.ac.uk/assa/2019/12/09/chinas-social-credit-system-the-chinese-citizens-perspective/.

8. "After 65-Yr-Old Starves to Death in Jharkhand, Authorities Release 50 Kg Cereal, Deny Family's Claims," *India Today*, June 7, 2019, https://www.indiatoday.in/india/story/65-yr-old-allegedly-starves-to-death-jharkhand-authorities-release-cereal-deny-family-s-claims-1544438-2019-06-07.

9. Apurva Vishwanath and Kaunain Sheriff, "Explained: What NRC+CAA Means to You," *Indian Express*, December 25, 2019, https://indianexpress.com/article/explained/explained-citizenship-amendment-act-nrc-caa-means-6180033/.

10. Jamiles Lartey, "By the Numbers: US Police Kill More in Days than Other Countries Do in Years," *The Guardian*, June 9, 2015, https://www.theguardian.com/us-news/2015/jun/09/the-counted-police-killings-us-vs-other-countries. Alexi Jones and Wendy Sawyer, "Not Just 'a Few Bad Apples': U.S. Police Kill Civilians at Much Higher Rates than Other Countries," *Prison Policy Initative* (blog), June 5, 2020, https://www.prisonpolicy.org/blog/2020/06/05/policekillings/.

11. United Nations Office of the High Commissioner on Human Rights, "World Stumbling Zombie-like into a Digital Welfare Dystopia, Warns UN Human Rights Expert," press release, October 17, 2019, https://www.ohchr.org/EN/NewsEvents/Pages/DisplayNews.aspx?NewsID=25156.

12. Aaron Moselle, "Algorithm to Reform Criminal Sentencing in Pa. Faces Deluge of Criticism," *All Things Considered*, December 13, 2018, https://whyy.org/articles/algorithm-to-reform-criminal-sentencing-in-pa-faces-deluge-of-criticism/.

13. One of Palantir's earliest and long-standing contracts is with the Centers for Medicare and Medicaid Services (CMS), for fraud detection.

14. Marisa Iati, "Florida Fired Its Coronavirus Data Scientist. Now She's Publishing the Statistics on Her Own," *Washington Post*, June 16, 2020, https://www.washingtonpost.com/nation/2020/06/12/rebekah-jones-florida-coronavirus/. In May 2021, Jones was given official "whistleblower" status by the Florida Office of the Attorney General. Sarah Blaskey, "Former Health Department Employee, Rebekah Jones, Granted Official Whistleblower Status," *Tampa Bay Times*, May 28, 2021, https://www.tampabay.com/news/health/2021/05/28/former-health-department-employee-rebekah-jones-granted-official-whistleblower-status/.

15. "The Belmont Report: Ethical Principles and Guidelines for the Protection of Human Subjects of Research,"April 18, 1979, can be read in full on the Department of Health and Human Services website at https://www.hhs.gov/ohrp/regulations-and-policy/belmont-report/read-the-belmont-report/index.html.

16. Western medicine's history of unilaterally demanding trust from patients is a very long one indeed.

17. The Tuskegee Syphilis Study, originally dubbed the "Tuskegee Study of Untreated Syphilis in the Negro Male," commenced in 1932. The research was not only executed by the federal government and the CDC, but also continued to be sanctioned by the medical establishment, including the American Medical Association. It was halted decades later, in 1972, after several whistleblowers went to the press about the inhumane basis of the study (Jones 1993).

18. American Public Health Association, "Racism Is a Public Health Crisis," American Public Health Association, June 2020, https://www.apha.org/topics-and-issues/health-equity/racism-and-health/racism-declarations. American Medical Association, "AMA Board of Trustees Pledges Action against Racism, Police Brutality," American Medical Association, June 7, 2020, https://www.ama-assn.org/press-center/ama-statements/ama-board-trustees-pledges-action-against-racism-police-brutality.

19. Isaac Stanley-Becker, "Trump's Promise of 'Warp Speed' Fuels Anti-Vaccine Movement in Fertile Corners of the Web" *Washington Post*, May 20, 2020, https://www.washingtonpost.com/national/trumps-promise-of-warp-speed-fuels-anti-vaccine-movement-in-fertile-corners-of-the-web/2020/05/20/c2b3d408-9ab2-11ea-89fd-28fb313d1886_story.html. Jan Hoffman, "Mistrust of a Coronavirus Vaccine Could Imperil Widespread Immunity," *New York Times*, July 18, 2020, https://www.nytimes.com/2020/07/18/health/coronavirus-anti-vaccine.html.

20. Quoted in Justin Petrone, "Study Investigates Trust in Genomic Data Sharing Initiatives to Inform Future Efforts," *GenomeWeb*, September 27, 2019, https://www.genomeweb.com/genetic-research/study-investigates-trust-genomic-data-sharing-initiatives-inform-future-efforts.

21. Dhruv Khullar, "Do You Trust the Medical Profession?" *New York Times*,

January 23, 2018, https://www.nytimes.com/2018/01/23/upshot/do-you-trust-the-medical-profession.html.

22. Ari Shapiro and Maureen Pao, "California and Texas Health Officials: Mistrust a Major Hurdle for Contact Tracers," *All Things Considered*, August 10, 2020, https://www.npr.org/sections/coronavirus-live-updates/2020/08/10/901064505/california-and-texas-health-officials-on-challenges-they-face-in-contact-tracing.

23. Shapiro and Pao, "California and Texas Health Officials."

24. Quoted from a slide presentation shared with Consortium members by Steve.

25. David Crary, "States Pushing Near-Bans on Abortion, Targeting Roe v. Wade," *Associated Press News*, April 10, 2019, https://apnews.com/3a9b3bc0e14d47aa8691aca84c32f391. These state-level bills are commonly called "abortion bans" because the legislation effectively makes all abortion illegal.

26. The term "honest broker" in the context of health informatics research is an official designation for the neutral third-party person or entity appointed by an ethics review board of a medical research institution to de-identify biospecimens and private information of research subjects. For multidisciplinary approaches on the use of honest brokers see Dhir et al. 2008.

27. HIPAA's subsequent amendments, such as the Privacy Rule, Security Rule, and the Omnibus Rule, gave more detail and direction on the proper handling and securing of digital health records and other data by covered entities. The right to give informed consent over what happens to a patient's data, however, was never legislated.

28. Lee Drutman and Steve Teles, "Why Congress Relies on Lobbyists Instead of Thinking for Itself," *The Atlantic*, March 10, 2015, https://www.theatlantic.com/politics/archive/2015/03/when-congress-cant-think-for-itself-it-turns-to-lobbyists/387295/. Cecilia Kang and Kenneth P. Vogel, "Tech Giants Amass a Lobbying Army for an Epic Washington Battle," *New York Times*, June 5, 2019, https://www.nytimes.com/2019/06/05/us/politics/amazon-apple-facebook-google-lobbying.html.

29. Kang and Vogel, "Tech Giants Amass a Lobbying Army."

30. Dipayan Ghosh, "What You Need to Know About California's New Data Privacy Law," *Harvard Business Review*, July 11, 2018, https://hbr.org/2018/07/what-you-need-to-know-about-californias-new-data-privacy-law.

31. SynthDoc is a pseudonym. In order to keep the company anonymous here, I have summarized the marketing language that appears on its website.

32. The Notice of Privacy Practices (NPP) is a boilerplate document that every healthcare provider is mandated to post or make available to patients under HIPAA. When a patient completes any intake documents, such as signing a document in which the patient consents to allowing the doctor to provide medical care, the patient will also be given a form to sign acknowledging that they were shown the practice's NPP. Patients are not consenting to their data being shared; they

don't have that right. Patients are simply acknowledging that their doctor's NPP was shown to them. Most NPPs are written in English (when I downloaded Ascension's NPP for their healthcare facilities in Alabama, the NPP was available only in English) and most have been found to have a Flesch Reading Ease score that is comprehensible to someone with post-PhD level education (Breese and Burman 2005).

33. In some research contexts, such as the European Union under the GDPR rules, research management companies have made consent a tradable asset as well (Beauvisage and Mellet 2020).

CHAPTER 3. COLLECTING LIFE

1. "Cardiologist Turns Promotional Pens into an Art Form," Cedars Sinai Newsroom, June 29, 2018, https://www.cedars-sinai.org/newsroom/cardiologist-turns-promotional-pens-into-an-art-form/.

2. In financial statements, such as the 10-K balance sheet report, which the Securities and Exchange Commission (SEC) requires that publicly traded corporations submit annually, a company's "goodwill" is accounted for on the assets side of the balance sheet. A company's goodwill is understood to be a nonphysical asset that nonetheless has a financial value.

3. Valentina Zarya, "Employers Are Quietly Using Big Data to Track Employee Pregnancies," *Fortune*, February 17, 2016, https://fortune.com/2016/02/17/castlight-pregnancy-data/.

4. Drew Harwell, "The Pregnancy-Tracking App Ovia Lets Women Record Their Most Sensitive Data for Themselves—and Their Boss," *Washington Post*, April 10, 2019, https://www.washingtonpost.com/technology/2019/04/10/tracking-your-pregnancy-an-app-may-be-more-public-than-you-think/.

5. The BRCA Sisterhood, an online support group of more than ten thousand women who are carriers of the BRCA gene markers, and are survivors or "previvors" of breast cancer, formed a patient data security advocacy group, the Light Collective in 2019, in light of the Cambridge Analytica breach of Facebook data. Kate Yandell, "Secure Connections: Patients Find Each Other Online and Get Support They Say Is Unparalleled, but with Openness Comes Concern about Privacy," *Cancer Today*, December 23, 2019, https://www.cancertodaymag.org/Pages/Winter2019-2020/Secure-Connections.aspx.

6. The whistleblower revealed that they work for Project Nightingale, the Google name for the cloud-based AI platform collaboration with Ascension. They initially tipped *Wall Street Journal* reporter Rob Copeland, who published the first story on November 11, 2019; Rob Copeland, "Google's 'Project Nightingale' Gathers Personal Health Data on Millions of Americans; Search Giant Is Amassing Health Records from Ascension Facilities in 21 States; Patients Not Yet Informed," *Wall Street Journal*, November 11, 2019. In subsequent reporting by the *New York*

Times and the *Guardian*, the nature of the secret partnership was corroborated and elaborated. Ed Pilkington, "Google's Secret Cache of Medical Data Includes Names and Full Details of Millions," *The Guardian*, November 12, 2019, https://www.theguardian.com/technology/2019/nov/12/google-medical-data-project-nightingale-secret-transfer-us-health-information. Kari Paul, "Google's Healthcare Partnership Sparks Fears for Privacy of Millions," *The Guardian*, November 12, 2019, https://www.theguardian.com/technology/2019/nov/11/google-healthcare-ascension-privacy-health-data. Natasha Singer and Daisuke Wakabayashi, "Google to Store and Analyze Millions of Health Records," *New York Times*, November 12, 2019.

7. One of the slides, towards the middle of the eight-minute video, issues a clear warning: "Confidential: Everything in this presentation is confidential, unless explicitly marked otherwise." The full video, uploaded November 12, 2019, was deleted by the video-sharing platform Dailymotion for a breach of its terms of service by the anonymous whistleblower. Presumably the leak itself was considered to be in violation of the platform's terms of service.

8. Anonymous, "I'm the Google Whistleblower. The Medical Data of Millions of Americans Is at Risk," *The Guardian*, November 14, 2019, https://www.theguardian.com/commentisfree/2019/nov/14/im-the-google-whistleblower-the-medical-data-of-millions-of-americans-is-at-risk.

9. Tariq Shakaut, "Our Partnership with Ascension," *Inside Google Cloud* (blog), November 12, 2019, https://cloud.google.com/blog/topics/inside-google-cloud/our-partnership-with-ascension/.

10. Madhumita Murgia, "NHS Trusts Sign First Deals with Google," *Financial Times*, September 18, 2019, https://www.ft.com/content/641e0d84-da21-11e9-8f9b-77216ebe1f17.

11. Health Information Technology for Economic and Clinical Health (HITECH) Act, 42 USC 201 (2009).

12. Dhruv Khullar, "The Unhealthy Politics of Pork: How It Increases Your Medical Costs," *New York Times*, October 28, 2017, https://www.nytimes.com/2017/10/25/upshot/the-unhealthy-politics-of-pork-how-it-increases-your-medical-costs.html.

13. Khullar, "The Unhealthy Politics of Pork."

14. Lee Drutman and Steve Teles, "Why Congress Relies on Lobbyists Instead of Thinking for Itself," *The Atlantic*, March 10, 2015, https://www.theatlantic.com/politics/archive/2015/03/when-congress-cant-think-for-itself-it-turns-to-lobbyists/387295/. Kimberly Kindy, "In Trump Era, Lobbyists Boldly Take Credit for Writing a Bill to Protect Their Industry," *Washington Post*, August 1, 2017, https://www.washingtonpost.com/powerpost/in-trump-era-lobbyists-boldly-take-credit-for-writing-a-bill-to-protect-their-industry/2017/07/31/eb299a7c-5c34-11e7-9fc6-c7ef4bc58d13_story.html.

15. Kathy Kemper, "Power to the Patients!" *Huffington Post*, January 8, 2018,

https://www.huffingtonpost.com/entry/power-to-the-patients_us_5a53fbede4bo f9b24bf31aoc.

16. Debra L. Ness, 2014. "Five Years after HITECH Act, Health IT Is Improving Care for Patients and Families," *HIT Consultant*, February 17, 2014, https://hitconsultant.net/2014/02/17/5-years-after-hitech-act-health-it-is-improving-care-for-patients/.

17. Snell, Elizabeth. 2018. "Cerner, Epic Systems Account for 51.5% Acute Care Hospital Market." *EHRIntelligence*, May 25, 2018. https://ehrintelligence.com/news/cerner-epic-systems-account-for-51.5-acute-care-hospital-market. Together Epic and Cernercomprise 51 percent of market share in the electronic health records industry; Tom Sullivan, "Epic, Cerner, Allscripts Moving to Dominate the Population Health Market," *HealthITNews*, June 7, 2017, https://www.healthcareitnews.com/news/epic-cerner-allscripts-moving-dominate-population-health-market.

18. Atul Gawande, "Why Doctors Hate Their Computers," *New Yorker*, November 5, 2018, https://www.newyorker.com/magazine/2018/11/12/why-doctors-hate-their-computers.

19. Kristen Hovet, "Your Genetic Data Is the New Oil. These Startups Will Pay to Rent It," *Leapsmag* (blog), September 21, 2018, https://leapsmag.com/your-genetic-data-is-the-new-oil-these-startups-will-pay-to-rent-it/.

20. Sorrell et al. v. IMS Health Inc., 564 U.S. 552 (2011).

21. "Payers" is the healthcare industry's term for health insurers and public sector entities, such as Veteran Affairs or CMS, who are responsible for bill remits. Patients who are uninsured are called self-payers. IQVIA was formally IMS Health Inc., the defendant in Sorrell v. IMS Health Inc. IQVIA's marketing tag line reads: "From evidence. To engagement. To the entire ecosystem." https://www.iqvia.com/.

22. John S. Toussaint, "Why Haven Healthcare Failed," *Harvard Business Review*, January 6, 2021, https://hbr.org/2021/01/why-haven-healthcare-failed.

23. Steve Inskeep, "What Do Amazon, JPMorgan Chase and Berkshire Hathaway Have Planned for Health Care?" *Morning Edition*, January 31, 2018, https://www.npr.org/2018/01/31/582056644/what-do-amazon-jpmorgan-chase-and-berkshire-hathway-have-planned-for-health-care.

24. H.R. 4613—Ensuring Patient Access to Healthcare Records Act of 2017, available at https://www.congress.gov/bill/115th-congress/house-bill/4613?s=1&r=57. Several months after the bill's passage in the US House of Representatives, I was contacted by a legislative assistant assisting a congressional member working on the House Committee on Energy and Commerce. The congressional member was concerned about the bill's sweeping language, and the power the company that drafted the bill increasingly holds over patient data. I promised anonymity to the legislative assistant who contacted me, as well as to the congressional member that the assistant works for. Additionally, I promised the assistant that I would not

discuss the identity of the corporation, one of the market leaders in the information and data services sector, that drafted H.R. 4613.

25. H.R. 4613—Ensuring Patient Access to Healthcare Records Act of 2017, 3.

26. H.R. 4613—Ensuring Patient Access to Healthcare Records Act of 2017, 4.

27. A 2019 study found that a majority of insured patients, 63 percent, don't use their online records portal. American Medical Association, "Most of Your Patients Still Aren't Using the Portal. Here's Why," *Health Affairs*, May 14, 2019, https://www.ama-assn.org/practice-management/digital/most-your-patients-still-aren-t-using-portal-here-s-why.

CHAPTER 4. MOBILIZING ALTERNATIVE DATA

1. Ron Marshall, "How Many Ads Do You See in One Day? Get Your Advertising Campaigns Heard," *Red Crow Marketing*, September 10, 2015, https://www.redcrowmarketing.com/2015/09/10/many-ads-see-one-day/.

2. Carole Cadwalladr and Emma Graham-Harrison, "Revealed: 50 Million Facebook Profiles Harvested for Cambridge Analytica in Major Data Breach," *The Guardian*, March 17, 2018, https://www.theguardian.com/news/2018/mar/17/cambridge-analytica-facebook-influence-us-election. Matthew Rosenberg, Nicholas Confessore, and Carole Cadwalladr, "How Trump Consultants Exploited the Facebook Data of Millions," *New York Times*, March 17, 2018, https://www.nytimes.com/2018/03/17/us/politics/cambridge-analytica-trump-campaign.html. Cecilia Kang and Sheera Frenkel, "Facebook Says Cambridge Analytica Harvested Data of Up to 87 Million Users," *New York Times*, April 4, 2018, https://www.nytimes.com/2018/04/04/technology/mark-zuckerberg-testify-congress.html.

3. Harry Davies, "Ted Cruz Campaign Using Firm That Harvested Data on Millions of Unwitting Facebook Users," *The Guardian*, December 11, 2015, https://www.theguardian.com/us-news/2015/dec/11/senator-ted-cruz-president-campaign-facebook-user-data.

4. On January 4, 2020, Brittany Kaiser, a second ex-Cambridge Analytica employee, who became a more significant and central whistleblower than the first, Christopher Wylie, uploaded five folders containing leaked documents from various election campaigns that she and others had worked on at Cambridge Analytica. The folders were labeled "Kenya," "Malaysia," "Brazil," "Iran," and "John Bolton," and represented five of the sixty-eight countries and political campaigns that the company had worked on before the Facebook data breach in 2018. Much of the leaked information had been submitted under subpoena to the Department of Justice's investigation, led by Robert Mueller, into whether the Trump Administration had colluded with the Russian government to illegally influence the 2016 US presidential election. The leaked files are no longer available online.

5. Carole Cadwalladr, "Fresh Cambridge Analytica Leak 'Shows Global Manipulation Is Out of Control,'" *The Observer*, January 4, 2020, https://www.thegua

rdian.com/uk-news/2020/jan/04/cambridge-analytica-data-leak-global-election-manipulation.

6. Thomas Germain, "Digital Billboards Are Tracking You. And They Really, Really Want You to See Their Ads," *Consumer Reports*, November 20, 2019, https://www.consumerreports.org/privacy/digital-billboards-are-tracking-you-and-they-want-you-to-see-their-ads/.

7. Curt Woodward and Hiawatha Bray, "Healey Halts Digital Ads Targeted at Women's Reproductive Clinics," *BostonGlobe*, April 4, 2017, https://www.bostonglobe.com/business/2017/04/04/healey-halts-digital-ads-targeted-women-reproductive-clinics/AoyPUG8u9hq9bJUAKC5gZN/story.html.

8. For more information on the CMS regulation, see "What Are the Value-Based Programs?" Centers for Medicare and Medicaid Service, https://www.cms.gov/Medicare/Quality-Initiatives-Patient-Assessment-Instruments/Value-Based-Programs/Value-Based-Programs, accessed December 2021.

9. The risk adjustment models, the CMS-Hierarchical Conditions Categories score, and the Health and Human Services (HHS) score are required in order for a health provider to be paid for medical services that are billed to Medicaid or Medicare or are billed to ACA Marketplace health insurers under the Affordable Care Act (Centers for Medicare and Medicaid Services 2020b).

10. The International Classification of Diseases (ICD) is managed by the World Health Organization and is in its eleventh revision. In the tenth revision, the number of clinical diagnostic codes jumped to sixty-eight thousand, compared with thirteen thousand in the ICD-9. Each country adapts the ICD to fit the specifications of its healthcare system needs. For the United States, the Centers for Medicare and Medicaid Services, which is the regulator in this case, has developed the ICD-10-CM for clinical diagnostic coding and the ICD-10-PCS, another schema of procedural codes, for the complex billing needs of inpatient care. The ICD-10-PCS schema contains eighty-seven thousand medical procedural codes.

11. LexisNexis. 2016. "LexisNexis Socioeconomic Data Coverage." For more information on the LexisNexis "socioeconomic health scores," visit https://risk.lexisnexis.com/products/socioeconomic-health-score.

12. The FICO Score, originally developed by the Fair Isaac Company (FICO) in 1989, has become a proprietary eponym for credit scores. See https://www.fico.com/25years/.

13. Dave Barkholz, "Equifax Breach Exposes Healthcare Vendor Vulnerabilities," *Modern Healthcare*, September 12, 2017, http://www.modernhealthcare.com/article/20170912/NEWS/170919966.

14. Omnibus Final Rule, *supra* note 5. A business associate can be any contractor or subcontractor doing business with or on behalf of a covered entity, such as an EHR platform vendor or an accounting firm, general counsel, et cetera. 45. C.F.R. 160.103 (Office of Civil Rights 2013). *See* Omnibus Final Rule, *supra* note 5.

For an explanation of the new oversight of business associates under the Omnibus Rule, see the HITECH Act (111th Congress 2009).

15. Stacy Cowley and Tara Siegel Bernard, "As Equifax Amassed Ever More Data, Safety Was a Sales Pitch," *New York Times*, September 23, 2017.

16. Experian Marketing Services, "Product Guide: Mosaic® USA," brochure, http://www.experian.com/assets/marketing-services/brochures/mosaic-brochure-october-2014.pdf.

17. Cowley and Bernard, "As Equifax Amassed Ever More Data."

18. Natasha Singer, "You for Sale: Mapping, and Sharing, the Consumer Genome," *New York Times*, June 17, 2012.

19. Janet Vertesi, "My Experiment Opting Out of Big Data Made Me Look Like a Criminal," *Time*, May 1, 2014, http://time.com/83200/privacy-internet-big-data-opt-out/.

20. For example, the Gramm Leach Bailey Act regulates the handling and disclosure of consumer nonpublic personal information by the banking and finance industry, at 15 U.S.C. §§6801-6809 (Federal Trade Commission 2000).

21. For a definition of protected health information, see 45 C.F.R. §160.103. In my previous work, especially in *Healthcare and Big Data* (2016) I have theorized that the PHI becomes a regulatory data object under HIPAA, a *thing* with material consequences, bestowed a special status with unique properties through classification by the legislative fiction of categorization.

22. 45 C.F.R. §160.103 (2002; 2013). In an interview with a health informatics researcher who regularly uses private health data in his role at a large, multinational professional services firm, he described PHI that had not been de-identified as "nuclear waste" and "radioactive" because the risks of heavy fines and other penalties to the third party in possession of identifiable health data were too great. Once these data were de-identified, however, they were considered safe to handle.

23. Homepage, IQVIA, https://www.iqvia.com/, accessed May, 2018. For a discussion of "detailing," see Sorrell v. IMS Health, 564 U.S. 552 (2011).

24. "Grindr Shared Information about Users' HIV Status with Third Parties," *The Guardian*, April 3, 2018, http://www.theguardian.com/technology/2018/apr/03/grindr-shared-information-about-users-hiv-status-with-third-parties.

25. Brian Krebs, "Experian Sold Consumer Data to ID Theft Service," *Krebson-Security* (blog), October 20, 2013, http://krebsonsecurity.com/2013/10/experian-sold-consumer-data-to-id-theft-service/.

26. "Equifax Announces Cybersecurity Incident Involving Consumer Information," Equifax Inc., September 7, 2017, https://investor.equifax.com/news-events/press-releases/detail/240/equifax-announces-cybersecurity-incident-involving-consumer.

27. "Equifax to Pay $575 Million as Part of Settlement with FTC, CFPB, and States Related to 2017 Data Breach," Federal Trade Commission, press release, July 22, 2019, https://www.ftc.gov/news-events/press-releases/2019/07/equifax

-pay-575-million-part-settlement-ftc-cfpb-states-related. Antonio Villas-Boas and Anaele Pelisson, "The Equifax Hack Isn't the Biggest Security Breach of All Time, but It Could Be One of the Worst in History for Americans," *Business Insider*, September 8, 2017, https://www.businessinsider.com/how-equifax-compares-to-biggest-hacks-of-all-time-chart-2017-9.

28. European Union, General Data Protection Regulation (GDPR), 2016, Official Legal Text, available at https://gdpr-info.eu/.

CHAPTER 5. ON SCORING LIFE

1. Carl, the data manager working for the regional data-sharing organization described in chapter 2, used this aphorism in an interview to describe the work that he does as a data scientist. The mantra has taken on various iterations, especially in business management and financial risk literature, such as "If you can't measure it, you can't improve it," or "You can't manage what you can't measure," both sayings apocryphally attributed to business management gurus Peter Drucker or W. Edwards Deming (Berenson 2016).

2. Sarah Kessler, "Companies Are Using Employee Survey Data to Predict—and Squash—Union Organizing," *OneZero*, July 30, 2020, https://onezero.medium.com/companies-are-using-employee-survey-data-to-predict-and-squash-union-organizing-a7e28a8c2158. Emma Goldberg, "Personality Tests Are the Astrology of the Office," *New York Times*, September 17, 2019, https://www.nytimes.com/2019/09/17/style/personality-tests-office.html.

3. William Heisel, "Why We Should Tread Carefully When Reporting on Adverse Childhood Experiences," Center for Health Journalism, *Children's Health Matters* (blog), July 1, 2019, https://centerforhealthjournalism.org/2019/06/03/why-we-should-tread-carefully-when-reporting-adverse-childhood-experiences. There is debate regarding premature mortality due to a high ACE score alone, as none of the graded ACES categories show an increase in the risk for early death (Brown et al. 2009).

4. Julia Angwin, Jeff Larson, Surya Mattu, and Lauren Kirchner, "Machine Bias," *ProPublica*, May 23, 2016, https://www.propublica.org/article/machine-bias-risk-assessments-in-criminal-sentencing.

5. Angwin et al., "Machine Bias." In a response to criticisms of the recidivism risk-scoring tool, research scientists Christina Mendoza and Eugenie Jackson at equivant/Northpointe Inc., the developers of COMPAS, acknowledge that while the tool may not have been adequately and independently validated in its earliest adoption by some jurisdictions in the mid-1990s, in the intervening years it has been tested and verified by five independent studies. They point out that the tool has also been refined over the years to account for complexity and to include new research and thinking about the factors that influence criminal behavior and the risk of reoffending (Jackson and Mendoza 2020).

6. The engineering team that developed the Allegheny Family Screening Tool have written and spoken publically about their efforts to make the model transparent and fair, and have expressed deep concern about how the platform is failing families. On the website for the risk score, it is described as "predictive modelling in child welfare in Alleghany County." More on the tool can be found at "The Alleghany Family Screening Tool," Alleghany County, Human Services, https://www.alleghenycounty.us/Human-Services/News-Events/Accomplishments/Allegheny-Family-Screening-Tool.aspx, accessed December 2021.

7. Jorge Luis Borges, "On Exactitude in Science." In *The Aleph and Other Stories,* edited by Andrew Hurley. New York: Penguin Classics, 2000, p. 181.

8. Marshall Allen, "Health Insurers Are Vacuuming Up Details About You—And It Could Raise Your Rates," *ProPublica,* July 17, 2018, https://www.propublica.org/article/health-insurers-are-vacuuming-up-details-about-you-and-it-could-raise-your-rates.

9. Allen, "Health Insurers Are Vacuuming Up Details About You."

10. Jeremy M. Simon, "New Medical FICO Score Sparks Controversy, Questions," *CreditCards.com,* July 28, 2011, https://www.creditcards.com/credit-card-news/fico-score-medication-adherence-1270.php.

11. Robert Berner and Chad Terhune, "Hospitals X-Ray Patient Credit Scores: More and More Are Buying Credit Data to See If the Sick Can Afford Treatment," *Bloomberg Businessweek,* November 20, 2008, http://tinyurl.com/DROMBerner.

12. Frank Pasquale, "Quantifying Love," *Boston Review,* Fall 2018, 106–113.

13. 116th Congress, "H.R. 3621: Comprehensive CREDIT Act of 2020," https://www.congress.gov/bill/116th-congress/house-bill/3621/text. In particular, Section 501, Consumer Bureau Study and Report to Congress on the Impact of Non-Traditional Data, and Section 502 § 631, Credit Scoring Models, address the use and validation of nontraditional models and data in consumer credit scoring, and making these scores more transparent.

14. Shannon Pettypiece and Jordan Robertson, "Hospitals Are Mining Patients' Credit Card Data to Predict Who Will Get Sick," *Bloomberg,* July 3, 2014, https://www.bloomberg.com/news/articles/2014-07-03/hospitals-are-mining-patients-credit-card-data-to-predict-who-will-get-sick.

15. Shelby Livingston, "UnitedHealthcare Expects Big Medicare Advantage Gains in 2020," *Modern Healthcare,* January 15, 2020, https://www.modernhealthcare.com/insurance/unitedhealthcare-expects-big-medicare-advantage-gains-2020.

CHAPTER 6. DATA VISIBILITIES

1. I was given permission to retell my friend's story here. I have used a pseudonym and changed some details about her life.

2. Janet Vertesi, "My Experiment Opting Out of Big Data Made Me Look Like

a Criminal," *Time*, May 1, 2014, http://time.com/83200/privacy-internet-big-data-opt-out/.

3. Recent books by Keeanga-Yamahtta Taylor (2019) and by Richard Rothstein (2018) demonstrate how racially based redlining and discriminatory practices by the real estate and mortgage lending industries persist more than fifty years after redlining was officially outlawed by the passage of the Fair Housing Act of 1968.

4. "Customer Acquisition," VisualDNA, https://www.visualdna.com/creditandrisk/visualdna-for-customer-acquisition/, accessed April 2021.

5. Experian commercials and consumer educational videos can be viewed on the Experian YouTube channel, https://www.youtube.com/channel/UCrd9d2_jNpHkfOvZ-IO7XXA, accessed May, 2021.

6. Experian Marketing Services, "Product Sheet: Social Media Analysis," https://www.experian.com/assets/marketing-services/product-sheets/social-media-analysis.pdf, 1; my emphasis.

7. Olga Khazan, "What Happens When You Don't Pay a Hospital Bill," *The Atlantic*, August 28, 2019, https://www.theatlantic.com/health/archive/2019/08/medical-bill-debt-collection/596914/.

EPILOGUE

1. David Gelles, "How Tech Billionaires Hack Their Taxes with a Philanthropic Loophole," *New York Times*, August 3, 2018, https://www.nytimes.com/2018/08/03/business/donor-advised-funds-tech-tax.html.

References

Al Dahdah, Marine. 2019. "From Evidence-Based to Market-Based MHealth: Itinerary of a Mobile (for) Development Project." *Science, Technology, and Human Values* 44 (6): 1048–67. https://doi.org/10.1177/0162243918824657.

Allen, Anita L. 2011. *Unpopular Privacy: What Must We Hide?* Oxford: Oxford University Press.

Allen, Timothy, and Colin DeYoung. 2016. "Personality Neuroscience and the Five-Factor Model." In *The Oxford Handbook of the Five Factor Model*, edited by Thomas A. Widiger. Oxford: Oxford University Press.

Alston, Philip. 2019. "Report of the Special Rapporteur on Extreme Poverty and Human Rights." A/74/48037. United Nations Office of the High Commissioner for Human Rights. https://www.ohchr.org/EN/NewsEvents/Pages/DisplayNews.aspx?NewsID=25156.

Amaro, Ramon. 2020. "Machine Diagnosis." *Open*, April 28, 2020. https://onlineopen.org/machine-diagnosis.

Amoore, Louise. 2020. *Cloud Ethics: Algorithms and the Attributes of Ourselves and Others*. Durham, NC: Duke University Press.

Appiah, Kwame Anthony. 2017. *As If: Idealization and Ideals*. Cambridge, MA: Harvard University Press.

Ard, B. J. 2015. "Confidentiality and the Problem of Third Parties: Protecting Reader Privacy in the Age of Intermediaries." *Yale Journal of Law and Technology* 16 (1). http://digitalcommons.law.yale.edu/yjolt/vol16/iss1/1.

Auxier, Brooke, Lee Rainie, Monica Anderson, Andrew Perrin, Madhu Kumar, and Erica Turner. 2019. "Americans and Privacy: Concerned, Confused and Feeling Lack of Control Over Their Personal Information." Pew Research Center, November 15, 2019. https://www.pewresearch.org/internet/2019/11/15/americans-and-privacy-concerned-confused-and-feeling-lack-of-control-over-their-personal-information/.

Bartlett, Robert, Adair Morse, Richard Stanton, and Nancy Wallace. 2018. "Consumer-Lending Discrimination in the Era of FinTech." National Bureau of Economic Research Working Paper, 25943. Cambridge, MA: National Bureau of Economic Research.

Beard, Martha Perine. 2010. "In-Depth: Reaching the Unbanked and Underbanked." *Central Banker*, Winter 2010: 6–7. https://www.stlouisfed.org/publications/central-banker/winter-2010/reaching-the-unbanked-and-underbanked.

Beauvisage, Thomas, and Kevin Mellet. 2020. "Is Consent a Relevant Model for Digital Market Regulation?" Paper presented at the Moralizing the Data Economy panel, 4S EASST 2020 Conference, August 18–21, 2020. https://convention2.allacademic.com/one/ssss/ssss20/.

Beer, David. 2019. *The Data Gaze*. London: SAGE Publications.

Benjamin, Ruha. 2016. "Informed Refusal: Toward a Justice-Based Bioethics." *Science, Technology, and Human Values* 41 (6): 967–90. https://doi.org/10.1177/0162243916656059

———. 2019. *Race after Technology: Abolitionist Tools for the New Jim Code*. Medford, MA: Polity.

Bennett, Jeannette N. 2017. "Credit Bureaus: The Record Keepers." *Page One Economics*, December 7, 2017.

Berenson, Robert A. 2016. "If You Can't Measure Performance, Can You Improve It?" *JAMA Forum*, January 13, 2016. https://newsatjama.jama.com/2016/01/13/jama-forum-if-you-cant-measure-performance-can-you-improve-it/.

Berlant, Lauren. 2011. *Cruel Optimism*. Durham, NC: Duke University Press.

Berliner, Lauren S., and Nora J. Kenworthy. 2017. "Producing a Worthy Illness: Personal Crowdfunding amidst Financial Crisis." *Social Science and Medicine* 187 (August): 233–42. https://doi.org/10.1016/j.socscimed.2017.02.008.

Beyer, Kirsten M. M., Yuhong Zhou, Kevin Matthews, Amin Bemanian, Purushottam W. Laud, and Ann B. Nattinger. 2016. "New Spatially Continuous Indices of Redlining and Racial Bias in Mortgage Lending: Links to Survival after Breast Cancer Diagnosis and Implications for Health Disparities Research." *Health and Place* 40 (July): 34–43. https://doi.org/10.1016/j.healthplace.2016.04.014.

Bhatia, Kiran Vinod, and Radhika Gajjala. 2020. "Examining Anti-CAA Protests at Shaheen Bagh: Muslim Women and Politics of the Hindu India." *International Journal of Communication*, November, 6286–304.

Bien, Jeffrey, and Vinay Prasad. 2016. "Future Jobs of FDA's Haematology-

Oncology Reviewers." *BMJ* 354 (September): i5055. https://doi.org/10.1136/bmj.i5055.

Birch, Kean. 2017. "Rethinking Value in the Bio-Economy." *Science, Technology and Human Values* 42 (3): 460–90. https://doi.org/10.1177/0162243916661633.

Birch, Kean, and Fabian Muniesa. 2020. "Introduction: Assetization and Technoscientific Capitalism." In *Assetization: Turning Things into Assets in Technoscientific Capitalism*, edited by Kean Birch and Fabian Muniesa, 1–43. Cambridge, MA: MIT Press.

Bivens, Josh. 2016. "Why Is Recovery Taking so Long—and Who's to Blame?" Economic Policy Institute, August 11, 2016.

Blendon, Robert J., John M. Benson, and Joachim O. Hero. 2014. "Public Trust in Physicians: U.S. Medicine in International Perspective." *New England Journal of Medicine* 371 (17): 1570–72. https://doi.org/10.1056/NEJMp1407373.

Borges, Jorge Luis. 2000. "On Exactitude in Science." In *The Aleph and Other Stories*, edited by Andrew Hurley, 181. New York: Penguin Classics.

Bouk, Dan. 2015. *How Our Days Became Numbered: Risk and the Rise of the Statistical Individual*. Chicago: University of Chicago Press.

———. 2017. "The History and Political Economy of Personal Data over the Last Two Centuries in Three Acts." *Osiris* 32 (1): 85–106.

Boumil, Marcia M., Kaitlyn Dunn, Nancy Ryan, and Katrina Clearwater. 2012. "Prescription Data Mining, Medical Privacy and the First Amendment: The U.S. Supreme Court in Sorrell v. IMS Health Inc." *Annals of Health Law* 21 (2): 447–91.

Box, George E. P. 1979. "Science and Statistics." *Journal of the American Statistical Association* 71 (356): 791–99.

Bradley, Nigel. 2007. *Marketing Research: Tools and Techniques*. Oxford: Oxford University Press.

Breese, Peter, and William Burman. 2005. "Readability of Notice of Privacy Forms Used by Major Health Care Institutions." *Journal of the American Medical Association* 293 (13): 1593–94.

Brevoort, Kenneth P., Philipp Grimm, and Michelle Kambara. 2015. "Credit Invisibles." CFPB Data Point. Washington DC: Consumer Financial Protection Bureau.

Bridges, Khiara M. 2011. "Privacy Rights and Public Families." *Harvard Journal of Law and Gender* 34 (1): 113–74.

Brighenti, Andrea Mubi. 2017. "The Visible: Element of the Social." *Frontiers in Sociology* 2. https://doi.org/10.3389/fsoc.2017.00017.

Brown, David W., Robert F. Anda, Henning Tiemeier, Vincent J. Felitti, Valerie J. Edwards, Janet B. Croft, and Wayne H. Giles. 2009. "Adverse Childhood Experiences and the Risk of Premature Mortality." *American Journal of*

Preventive Medicine 37 (5): 389–96. https://doi.org/10.1016/j.amepre.2009.06.021.

Browne, Simone. 2015. *Dark Matters: On the Surveillance of Blackness*. Durham, NC: Duke University Press.

Buolamwini, Joy, and Timnit Gebru. 2018. "Gender Shades: Intersectional Accuracy Disparities in Commercial Gender Classification." In *Proceedings of Machine Learning Research*, 81:77–91. PMLR. http://proceedings.mlr.press/v81/buolamwini18a.html.

Burwell, Sylvia M., Steven VanRoekel, Todd Park, and Dominic J. Mancini. 2013."Memorandum for the Heads of Executive Departments and Agencies: Open Data Policy: Managing Information as an Asset," Executive Office of the President, Office of Management and Budget, memorandum, May 9, 2013. https://www.whitehouse.gov/sites/whitehouse.gov/files/omb/memoranda/2013/m-13-13.pdf.

Campolo, Alexander, and Kate Crawford. 2020. "Enchanted Determinism: Power without Responsibility in Artificial Intelligence." *Engaging Science, Technology, and Society* 6 (January): 1–19. https://doi.org/10.17351/ests2020.277.

Caplan, Arthur L. 1992. "When Evil Intrudes." *The Hastings Center Report* 22 (6): 29–32. https://doi.org/10.2307/3562946.

Centers for Medicare and Medicaid Services (CMS). n.d. "National Health Expenditure Data." https://www.cms.gov/Research-Statistics-Data-and-Systems/Statistics-Trends-and-Reports/NationalHealthExpendData, last modified December 1, 2021.

——. 2020. "Risk Adjustment in Quality Measurement." Supplemental Material, version 16.0. *CMS Measures Management System Blueprint*. Washington DC: Centers for Medicare and Medicaid Services.

Cheney-Lippold, John. 2011. "A New Algorithmic Identity: Soft Biopolitics and the Modulation of Control." *Theory, Culture and Society* 28 (6): 164–81. https://doi.org/10.1177/0263276411424420.

——. 2017. *We Are Data: Algorithms and the Making of Our Digital Selves*. New York: New York University Press.

Chico, Victoria. 2018. "The Impact of the General Data Protection Regulation on Health Research." *British Medical Bulletin* 128 (1): 109–18. https://doi.org/10.1093/bmb/ldy038.

Chouldechova, Alexandra. 2020. "Transparency and Simplicity in Criminal Risk Assessment." *Harvard Data Science Review* 2 (1). https://doi.org/10.1162/99608f92.b9343eec.

Chouldechova, Alexandra, Diana Benavides-Prado, Oleksandr Fialko, and Rhema Vaithianathan. 2018. "A Case Study of Algorithm-Assisted Decision Making in Child Maltreatment Hotline Screening Decisions." Proceedings of the first Conference on Fairness, Accountability and Transparency. *Proceed-*

ings of Machine Learning Research, 81:134–48. http://proceedings.mlr.press/v81/chouldechova18a.html.

Chun, Wendy Hui Kyong, Matthew Fuller, Lev Manovich, and Noah Wardrip-Fruin. 2011. *Programmed Visions: Software and Memory*. Cambridge, MA: MIT Press.

Citron, Danielle Keats, and Frank Pasquale. 2014. "The Scored Society: Due Process for Automated Predictions." *Washington Law Review* 89 (1): 1–33.

Cohen, Andrew J., Hartley Brody, German Patino, Medina Ndoye, Aron Liaw, Christi Butler, and Benjamin N. Breyer. 2019. "Use of an Online Crowdfunding Platform for Unmet Financial Obligations in Cancer Care." *JAMA Internal Medicine* 179 (12): 1717–20. https://doi.org/10.1001/jamainternmed.2019.3330.

Cohen-Cole, Ethan. 2010. "Credit Card Redlining." *The Review of Economics and Statistics* 93 (2): 700–713. https://doi.org/10.1162/REST_a_00052.

Consumer Finance Protection Bureau. 2012. "Key Dimensions and Processes in the U.S. Credit Reporting System: A Review of How the Nation's Largest Credit Bureaus Manage Consumer Data." Washington DC: Consumer Financial Protection Bureau. http://files.consumerfinance.gov/f/201212_cfpb_credit-reporting-white-paper.pdf.

———. 2014. "Consumer Credit Reports: A Study of Medical and Non-Medical Collections." Washington DC: Consumer Financial Protection Bureau. https://files.consumerfinance.gov/f/201412_cfpb_reports_consumer-credit-medical-and-non-medical-collections.pdf.

———. 2015. "Consumer Voices on Credit Reports and Scores." Washington DC: Consumer Financial Protection Bureau.

———. 2017. "Consumer Experiences with Debt Collection: Findings from the CFPB's Survey of Consumer Views on Debt." Washington DC: Consumer Financial Protection Bureau.

Cooper, Zack, Amanda E. Kowalski, Eleanor N. Powell, and Jennifer Wu. 2017. "Politics, Hospital Behavior, and Health Care Spending." National Bureau of Economic Research. Working Paper 23748. https://doi.org/10.3386/w23748.

Couldry, Nick, and Ulises A. Mejias. 2019. "Data Colonialism: Rethinking Big Data's Relation to the Contemporary Subject." *Television and New Media* 20 (4): 336–49. https://doi.org/10.1177/1527476418796632.

Couldry, Nick, and Jun Yu. 2018. "Deconstructing Datafication's Brave New World." *New Media and Society*, May 19, 2018. https://doi.org/10.1177/1461444818775968.

Courtland, Rachel. 2018. "Bias Detectives: The Researchers Striving to Make Algorithms Fair." *Nature* 558 (7710): 357–60. https://doi.org/10.1038/d41586-018-05469-3.

Craig, Jessica M., Chad R. Trulson, Matt DeLisi, and Jon W. Caudill. 2020. "Toward an Understanding of the Impact of Adverse Childhood Experiences on the Recidivism of Serious Juvenile Offenders." *American Journal of Criminal Justice* 45 (6): 1024–39. https://doi.org/10.1007/s12103-020-09524-6.

Crawford, Kate. 2021. *Atlas of AI: Power, Politics, and the Planetary Costs of Artificial Intelligence*. New Haven, CT: Yale University Press.

Crook, Errol D., and Mosha Peters. 2008. "Health Disparities in Chronic Diseases: Where the Money Is." *The American Journal of the Medical Sciences* 335 (4): 266–70. https://doi.org/10.1097/MAJ.0b013e31816902f1.

Daston, Lorraine, and Peter Galison. 1992. "The Image of Objectivity." *Representations*, no. 40 (Fall): 48.

Dean, Lorraine T., and Lauren H. Nicholas. 2018. "Using Credit Scores to Understand Predictors and Consequences of Disease." *American Journal of Public Health* 108 (111): 1503–5.

Dean, Lorraine T., Kathryn H. Schmitz, Kevin D. Frick, Lauren H. Nicholas, Yuehan Zhang, S. V. Subramanian, and Kala Visvanathan. 2018. "Consumer Credit as a Novel Marker for Economic Burden and Health after Cancer in a Diverse Population of Breast Cancer Survivors in the USA." *Journal of Cancer Survivorship* 12 (3): 306–15. https://doi.org/10.1007/s11764-017-0669-1.

Debt Collective, and Astra Taylor. 2020. *Can't Pay, Won't Pay: The Case for Economic Disobedience and Debt Abolition*. Chicago: Haymarket Books.

Deleuze, Gilles. 1992. "Postscript on the Societies of Control." *October* 59 (Winter): 3–7.

Delgado, M. Kit, Michael A. Yokell, Kristan L. Staudenmayer, David A. Spain, Tina Hernandez-Boussard, and N. Ewen Wang. 2014. "Factors Associated with the Disposition of Severely Injured Patients Initially Seen at Non-Trauma Center Emergency Departments: Disparities by Insurance Status." *JAMA Surgery* 149 (5): 422–30. https://doi.org/10.1001/jamasurg.2013.4398.

Department of Health and Human Services. 2002. "Standards for Privacy of Individually Identifiable Health Information; Final Rule." 45 CFR 160 and 164. https://www.federalregister.gov/documents/2002/08/14/02-20554/standards-for-privacy-of-individually-identifiable-health-information.

———. 2009. "Federal Policy for the Protection of Human Subjects ('Common Rule')." Office for Human Resource Protections. June 23, 2009. https://www.hhs.gov/ohrp/regulations-and-policy/regulations/common-rule/index.html.

Devakumar, Delan, Sujitha Selvarajah, Geordan Shannon, Kui Muraya, Sarah Lasoye, Susanna Corona, Yin Paradies, Ibrahim Abubakar, and E. Tendayi Achiume. 2020. "Racism, the Public Health Crisis We Can No Longer Ignore." *The Lancet* 395 (10242): e112–13. https://doi.org/10.1016/S0140-6736(20)31371-4.

Dhir, Rajiv, Ashok A. Patel, Sharon Winters, Michelle Bisceglia, Dennis Swanson, Roger Aamodt, and Michael J. Becich. 2008. "A Multi-Disciplinary Approach to Honest Broker Services for Tissue Banks and Clinical Data: A

Pragmatic and Practical Model." *Cancer* 113 (7): 1705–15. https://doi.org/10.1002/cncr.23768.

Dixon, Pam. 2013. "Congressional Testimony: What Information Do Data Brokers Have on Consumers, and How Do They Use It?" *World Privacy Forum*, December 18, 2013. https://www.worldprivacyforum.org/2013/12/testimony-what-information-do-data-brokers-have-on-consumers/.

Dixon-Román, Ezekiel. 2016. "Algo-Ritmo: More-Than-Human Performative Acts and the Racializing Assemblages of Algorithmic Architectures." *Cultural Studies ↔ Critical Methodologies* 16 (5): 482–90. https://doi.org/10.1177/1532708616655769.

Doyle, Charles. 2011. *A Dictionary of Marketing*. 3rd ed. Oxford: Oxford University Press.

Dressel, Julia, and Hany Farid. 2018. "The Accuracy, Fairness, and Limits of Predicting Recidivism." *Science Advances* 4 (1): eaao5580. https://doi.org/10.1126/sciadv.aao5580.

Du Bois, W. E. B. 2007. *The Philadelphia Negro: A Social Study*. The Oxford W. E. B. Du Bois, edited by Henry Louis Gates, Jr. Oxford: Oxford University Press.

Dullabh, Prashila, Adil Moiduddin, Christine Nye, and Lindsay Virost. 2011. "The Evolution of the State Health Information Exchange Cooperative Agreement Program: State Plans to Enable Robust HIE." Office of the National Coordinator for Health Information Technology, US Department of Health and Human Services. Bethseda, MD: NORC.

Economic Policy Institute. 2019. "The Productivity–Pay Gap." Washington DC: Economic Policy Institute. https://www.epi.org/productivity-pay-gap/.

Ebeling, Mary F. E. 2016. *Healthcare and Big Data: Digital Specters and Phantom Objects*. New York: Palgrave Macmillan.

———. 2018. "Uncanny Commodities: Policy and Compliance Implications for the Trade in Debt and Health Data." *Annals of Health Law and Life Sciences* 27 (2): 125–47.

———. 2020. "Who Gains from Our Online Lives?" *Current History* 119 (813): 37–39. https://doi.org/10.1525/curh.2020.119.813.37.

Ellis Neyra, Ren. 2020. *Cry of the Senses: Listening to Latinx and Caribbean Poetics*. Durham, NC: Duke University Press.

Engelmann, Severin, Mo Chen, Felix Fischer, Ching-yu Kao, and Jens Grossklags. 2019. "Clear Sanctions, Vague Rewards: How China's Social Credit System Currently Defines 'Good' and 'Bad' Behavior." In *Proceedings of the Conference on Fairness, Accountability, and Transparency*, 69–78. Atlanta: ACM. https://doi.org/10.1145/3287560.3287585.

Equifax Predictive Sciences. 2005. "Utilizing Credit Scoring to Predict Patient Outcomes." White Paper. Equifax Predictive Sciences Research Paper. Atlanta: Equifax Inc. https://docplayer.net/8204509-Utilizing-credit-scoring-to-predi

ct-patient-outcomes-an-equifax-predictive-sciences-research-paper-september-2005.html.

Etzioni, Amitai. 1999. *The Limits of Privacy*. New York: Basic Books.

Eubanks, Virginia. 2018. *Automating Inequality: How High-Tech Tools Profile, Police, and Punish the Poor*. New York: St. Martin's Press.

Fair Isaac Corporation. 2012. "From Big Data to Big Marketing: Seven Essentials." *FICO® Insights*, white paper 63. San Jose, CA: Fair Isaac Corporation.

Farid, Hany. 2018. "Digital Forensics in a Post-Truth Age." *Forensic Science International* 289 (August): 268–69. https://doi.org/10.1016/j.forsciint.2018.05.047.

Federal Reserve Bank of New York. 2021. "Quarterly Report on Household Debt and Credit: 2020: Q4." Center for Microeconomic Data, Federal Reserve Bank of New York. https://www.newyorkfed.org/medialibrary/interactives/householdcredit/data/pdf/hhdc_2020q4.pdf.

Federal Trade Commission. 2002. "In Brief: The Financial Privacy Requirements of the Gramm-Leach-Bliley Act." Policy Brief BUS53. Washington DC: Federal Trade Commission. https://www.ftc.gov/tips-advice/business-center/privacy-and-security/gramm-leach-bliley-act.

Feiner, John R., John W. Severinghaus, and Philip E. Bickler. 2007. "Dark Skin Decreases the Accuracy of Pulse Oximeters at Low Oxygen Saturation: The Effects of Oximeter Probe Type and Gender." *Anesthesia and Analgesia* 105 (6 Suppl): S18–23. https://doi.org/10.1213/01.ane.0000285988.35174.d9.

Felitti, Vincent J., Robert F. Anda, Dale Nordenberg, David F. Williamson, Alison M. Spitz, Valerie Edwards, Mary P. Koss, and James S. Marks. 1998. "Relationship of Childhood Abuse and Household Dysfunction to Many of the Leading Causes of Death in Adults: The Adverse Childhood Experiences (ACE) Study." *American Journal of Preventive Medicine* 14 (4): 245–58. https://doi.org/10.1016/S0749-3797(98)00017-8.

Flaxman, Penny. 2015. "The 10-Minute Appointment." *The British Journal of General Practice* 65 (640): 573–74. https://doi.org/10.3399/bjgp15X687313.

Florence, Curtis, Jonathan Shepherd, Iain Brennan, and Thomas Simon. 2011. "Effectiveness of Anonymised Information Sharing and Use in Health Service, Police, and Local Government Partnership for Preventing Violence Related Injury: Experimental Study and Time Series Analysis." *The BMJ* 342 (June). https://doi.org/10.1136/bmj.d3313.

Florida, Richard. 2014. *The Rise of the Creative Class*. Rev. ed. New York: Basic Books.

Floridi, Luciano. 2005. "The Ontological Interpretation of Informational Privacy." *Ethics and Information Technology* 7 (4): 185–200. https://doi.org/10.1007/s10676-006-0001-7.

Foucault, Michel. 1982. *The Archaeology of Knowledge*. New York: Vintage.

———. 2003. *The Birth of the Clinic*. 3rd ed. London: Routledge.

Fourcade, Marion. 2017. "The Fly and the Cookie: Alignment and Unhingement in Twenty-First-Century Capitalism." *Socio-Economic Review* 15 (3): 661–78. https://doi.org/10.1093/ser/mwx032.

Fourcade, Marion, and Kieran Healy. 2013. "Classification Situations: Life-Chances in the Neoliberal Era." *Accounting, Organizations and Society* 38 (8): 559–72. https://doi.org/10.1016/j.aos.2013.11.002.

———. 2016. "Seeing like a Market." *Socio-Economic Review* 15 (1): 9–29. https://doi.org/10.1093/ser/mww033.

Gallup Inc. n.d. "Confidence in Institutions." Gallup Historical Trends. https://news.gallup.com/poll/1597/Confidence-Institutions.aspx.

Garrett, Brandon, and John Monahan. 2019. "Assessing Risk: The Use of Risk Assessment in Sentencing." *Judicature* 103 (2). https://judicature.duke.edu/articles/assessing-risk-the-use-of-risk-assessment-in-sentencing/.

Garrison, Nanibaa' A., Nila A. Sathe, Armand H. Matheny Antommaria, Ingrid A. Holm, Saskia C. Sanderson, Maureen E. Smith, Melissa L. McPheeters, and Ellen W. Clayton. 2016. "A Systematic Literature Review of Individuals' Perspectives on Broad Consent and Data Sharing in the United States." *Genetics in Medicine* 18 (7): 663–71. https://doi.org/10.1038/gim.2015.138.

Garvie, Clare, Alvaro Bedoya, and Jonathan Frankle. 2016. *The Perpetual Line-Up: Unregulated Police Face Recognition in America*. Georgetown Law Center on Privacy and Technology Report, October 18, 2016. Washington DC: Georgetown Law Center. https://www.perpetuallineup.org/.

Gebru, Timnit, Jamie Morgenstern, Briana Vecchione, Jennifer Wortman Vaughan, Hanna Wallach, Hal Daumé III, and Kate Crawford. 2020. "Datasheets for Datasets." *ArXiv* 1803.09010 [Cs], March. http://arxiv.org/abs/1803.09010.

Gewin, Virginia. 2016. "Data Sharing: An Open Mind on Open Data." *Nature* 529 (7584): 117–19. https://doi.org/10.1038/nj7584-117a.

Glissant, Édouard. 1997. *Poetics of Relation*. Translated by Betsy Wing. Ann Arbor: University of Michigan Press.

Golbeck, Jennifer. 2009. *Computing with Social Trust*. Human-Computer Interaction Series. London: Springer.

Goodman, Melody. 2016. "White Fear Creates White Spaces and Exacerbates Health Disparities." *Institute for Public Health* (blog), February 16, 2016. https://publichealth.wustl.edu/white-fear-creates-white-spaces-and-exacerbates-health-disparities/.

Gordon, Colin. 2008. *Mapping Decline*. Philadelphia: University of Pennsylvania Press.

Graeber, David. 2014. *Debt: The First Five Thousand Years*. Rev. ed. Brooklyn, NY: Melville House.

———. 2015. *The Utopia of Rules: On Technology, Stupidity and the Secret Joys of Bureaucracy*. Brooklyn, NY: Melville House.

Grier, Sonya, and Carol A. Bryant. 2005. "Social Marketing in Public Health." *Annual Review of Public Health* 26: 319–35.

Grundy, Quinn, Kellia Chiu, Fabian Held, Andrea Continella, Lisa Bero, and Ralph Holz. 2019. "Data Sharing Practices of Medicines Related Apps and the Mobile Ecosystem: Traffic, Content, and Network Analysis." *BMJ* 364 (March). https://doi.org/10.1136/bmj.l920.

Hahn, Heather, Rayanne Hawkins, Alexander Carther, and Alena Stern. 2020. "Access for All: Innovation for Equitable SNAP Delivery." Center on Labor, Human Services, and Population. Washington DC: The Urban Institute.

Hall, Mark A, Fabian Camacho, Elizabeth Dugan, and Rajesh Balkrishnan. 2002. "Trust in the Medical Profession: Conceptual and Measurement Issues." *Health Services Research* 37 (5): 1419–39. https://doi.org/10.1111/1475-6773.01070.

Halpern, Orit. 2014. *Beautiful Data: A History of Vision and Reason since 1945*. Durham, NC: Duke University Press.

Hammond, Wizdom Powell. 2010. "Psychosocial Correlates of Medical Mistrust among African American Men." *American Journal of Community Psychology* 45 (1–2): 87–106. https://doi.org/10.1007/s10464-009-9280-6.

Harari, Yuval Noah. 2017. *Homo Deus: A Brief History of Tomorrow*. New York: HarperCollins.

Hartman, Saidiya. 2008a. *Lose Your Mother*. New York: Farrar, Straus and Giroux.

———. 2008b. "Venus in Two Acts." *Small Axe: A Caribbean Journal of Criticism* 12 (2): 1–14. https://doi.org/10.1215/-12-2-1.

Hayles, N. Katherine. 1999. *How We Became Posthuman: Virtual Bodies in Cybernetics, Literature, and Informatics*. Chicago: University of Chicago Press.

Himmelstein, David U., Robert M. Lawless, Deborah Thorne, Pamela Foohey, and Steffie Woolhandler. 2019. "Medical Bankruptcy: Still Common Despite the Affordable Care Act." *American Journal of Public Health* 109 (3): 431–33. https://doi.org/10.2105/AJPH.2018.304901.

Himmelstein, David U., Deborah Thorne, Elizabeth Warren, and Steffie Woolhandler. 2009. "Medical Bankruptcy in the United States, 2007: Results of a National Study." *The American Journal of Medicine* 122 (8): 741–46. https://doi.org/10.1016/j.amjmed.2009.04.012.

Hoffman, Kelly M., Sophie Trawalter, Jordan R. Axt, and M. Norman Oliver. 2016. "Racial Bias in Pain Assessment and Treatment Recommendations, and False Beliefs about Biological Differences between Blacks and Whites." *Proceedings of the National Academy of Sciences of the United States of America* 113 (16): 4296–301. https://doi.org/10.1073/pnas.1516047113.

Hornblum, Allen M. 1999. *Acres of Skin: Human Experiments at Holmesburg Prison*. New York: Routledge.

Hou, Vivian, Daniel Wood, JoElla Carman, George Railean, and Christina Baird. 2021. "Debt in America: An Interactive Map." Washington DC: The Urban Institute. https://apps.urban.org/features/debt-interactive-map.

Hurley, Mikella, and Julius Adebayo. 2016. "Credit Scoring in the Era of Big Data." *Yale Journal of Law and Technology* 18: 148–216.

Hyman, Louis. 2011. *Debtor Nation: The History of America in Red Ink*. Princeton, NJ: Princeton University Press.

———. 2012. "The Politics of Consumer Debt." *The ANNALS of the American Academy of Political and Social Science* 644 (1): 40–49. https://doi.org/10.1177/0002716212452721.

Ipsos MORI. 2017. "The One-Way Mirror: Public Attitudes to Commercial Access to Health Data." Wellcome Trust. November 20, 2017. https://doi.org/10.6084/m9.figshare.5616448.v1.

Jackson, Eugenie, and Christina Mendoza. 2020. "Setting the Record Straight: What the COMPAS Core Risk and Need Assessment Is and Is Not." *Harvard Data Science Review* 2 (1). https://doi.org/10.1162/99608f92.1b3dadaa.

Jacoby, Sara F., Beidi Dong, Jessica H. Beard, Douglas J. Wiebe, and Christopher N. Morrison. 2018. "The Enduring Impact of Historical and Structural Racism on Urban Violence in Philadelphia." *Social Science and Medicine* 199 (February): 87–95. https://doi.org/10.1016/j.socscimed.2017.05.038.

Jasanoff, Sheila. 2011. "Constitutional Moments in Governing Science and Technology." *Science and Engineering Ethics* 17 (4): 621–38.

Johnson, John M., and Andrew Melnikov. 2009. The Wisdom of Distrust: Reflections on Ukrainian Society and Sociology. In *Studies in Symbolic Interaction*, edited by Norman K. Denzin, 33:9–18. https://doi.org/10.1108/S0163-2396(2009)0000033003.

Jones, James H. 1993. *Bad Blood: The Tuskegee Syphilis Experiment*. Rev. ed. New York: Free Press.

Joynt Maddox, Karen E., Mat Reidhead, Jianhui Hu, Amy J. H. Kind, Alan M. Zaslavsky, Elna M. Nagasako, and David R. Nerenz. 2019. "Adjusting for Social Risk Factors Impacts Performance and Penalties in the Hospital Readmissions Reduction Program." *Health Services Research* 54 (2): 327–36. https://doi.org/10.1111/1475-6773.13133.

Keisler-Starkey, Katherine, and Lisa N. Bunch. 2020. "Health Insurance Coverage in the United States: 2019." Current Population Reports, P60-271. Washington DC: United States Census Bureau.

Khullar, Dhruv. 2019. "Building Trust in Health Care—Why, Where, and How." *JAMA* 322 (6): 507–9. https://doi.org/10.1001/jama.2019.4892.

Kim, Pauline T. 2017. "Data-Driven Discrimination at Work." *William and Mary Law Review* 58 (3): 857–936.

Kiviat, Barbara. 2019. "Credit Scoring in the United States." *Economic Sociology: The European Electronic Newsletter* 21 (1): 33–42.

Knapp, Emily A., and Lorraine T. Dean. 2018. "Consumer Credit Scores as a Novel Tool for Identifying Health in Urban U.S. Neighborhoods." *Annals of Epidemiology* 28 (10): 724–29.

Knowles, Bran, and John T. Richards. 2021. "The Sanction of Authority: Promoting Public Trust in AI." In *Proceedings of the 2021 ACM Conference on Fairness, Accountability, and Transparency*, 262–71. New York: Association for Computing Machinery. https://doi.org/10.1145/3442188.3445890.

Landecker, Hannah. 2000. "Immortality, In Vitro: A History of the HeLa Cell Line." In *Biotechnology and Culture: Bodies, Anxieties, Ethics*, edited by Paul Browdin, 53–72. Bloomington: Indiana University Press.

Latour, Bruno. 1987. *Science in Action: How to Follow Scientists and Engineers through Society*. Cambridge, MA: Harvard University Press.

Lauer, Josh. 2017. *Creditworthy: A History of Consumer Surveillance and Financial Identity in America*. New York: Columbia University Press.

Law, John. 2004. *After Method: Mess in Social Science Research*. London: Routledge.

Lazzarato, Maurizio. 2014. *Signs and Machines: Capitalism and the Production of Subjectivity*. Translated by Joshua David Jordan. Los Angeles: Semiotext(e).

———. 2015. *Governing by Debt*. Cambridge, MA: Semiotext(e).

Lebowitz, Fran. 2011. *The Fran Lebowitz Reader*. New York: Knopf Doubleday.

Leonelli, Sabina. 2016a. *Data-Centric Biology: A Philosophical Study*. Chicago: University of Chicago Press.

———. 2016b. "Open Data: Curation Is Under-Resourced." *Nature* 538 (41). https://doi.org/10.1038/538041d.

———. 2019. "Data—from Objects to Assets." *Nature* 574:317–20. https://doi.org/10.1038/d41586-019-03062-w.

Littwin, Angela. 2013. "Escaping Battered Credit: A Proposal for Repairing Credit Reports Damaged by Domestic Violence." *University of Pennsylvania Law Review* 161 (2): 363–429.

Mabry, Charles D. 2014. "Does a 'Wallet Biopsy' Lead to Inappropriate Trauma Patient Care?" *JAMA Surgery* 149 (5): 430–31. https://doi.org/10.1001/jamasurg.2013.4403.

Madden, Mary, Michele Gilman, Karen Levy, and Alice Marwick. 2017. "Privacy, Poverty, and Big Data: A Matrix of Vulnerabilities for Poor Americans." *Washington University Law Review* 95:53–125.

Mai, Jens-Erik. 2016. "Big Data Privacy: The Datafication of Personal Information." *The Information Society* 32 (3): 192–99. https://doi.org/10.1080/01972243.2016.1153010.

Manovich, Lev. 2012. "Trending: The Promises and the Challenges of Big Social Data." In *Debates in the Digital Humanities*, edited by Matthew K. Gold and Lauren F. Klein, 460–75. Minneapolis: University of Minnesota Press.

Mariotto, Angela B., K. Robin Yabroff, Yongwu Shao, Eric J. Feuer, and Martin L. Brown. 2011. "Projections of the Cost of Cancer Care in the United States: 2010–2020." *Journal of the National Cancer Institute* 103 (2): 117–28. https://doi.org/10.1093/jnci/djq495.

Marx, Karl. 1976. *Capital: A Critique of Political Economy*. Vol. 1. London: Penguin.

Mayer-Schönberger, Viktor, and Kenneth Cukier. 2012. *Big Data: A Revolution That Transforms How We Work, Live, and Think*. Boston: Houghton Mifflin Harcourt.

McDonald, Ken, and Theodore Murphy. 2015. "Predict, Protect, Prevent: Working toward a Personalized Approach to Heart Failure Prevention." *JACC Heart Failure* 3 (6): 456–58. https://doi.org/10.1016/j.jchf.2015.01.011.

Meister, Sven, Wolfgang Deiters, and Stefan Becker. 2016. "Digital Health and Digital Biomarkers—Enabling Value Chains on Health Data." *Current Directions in Biomedical Engineering* 2 (1): 577–81. https://doi.org/10.1515/cdbme-2016-0128.

Mellet, Kevin, and Thomas Beauvisage. 2019. "Cookie Monsters: Anatomy of a Digital Market Infrastructure." *Consumption Markets and Culture* 22 (5–6): 1–20. https://doi.org/10.1080/10253866.2019.1661246.

Merchant, Carolyn. 1990. *The Death of Nature: Women, Ecology, and the Scientific Revolution*. New York: HarperOne.

———. 2006. "The Scientific Revolution and The Death of Nature." *ISIS* 97: 513–33.

———. 2008. "'The Violence of Impediments': Francis Bacon and the Origins of Experimentation." *ISIS* 99: 731–60.

Meyer, Samantha, Paul Ward, John Coveney, and Wendy Rogers. 2008. "Trust in the Health System: An Analysis and Extension of the Social Theories of Giddens and Luhmann." *Health Sociology Review* 17 (2): 177–86. https://doi.org/10.5172/hesr.451.17.2.177.

Mijente. 2019. "The War against Immigrants: Trump's Tech Tools Powered by Palantir." Mijente, August 2019. https://mijente.net/wp-content/uploads/2019/08/Mijente-The-War-Against-Immigrants_-Trumps-Tech-Tools-Powered-by-Palantir_.pdf.

Milne, Richard, Katherine I. Morley, Heidi Howard, Emilia Niemiec, Dianne Nicol, Christine Critchley, Barbara Prainsack, et al. 2019. "Trust in Genomic Data Sharing among Members of the General Public in the UK, USA, Canada and Australia." *Human Genetics* 138:1237–46. https://doi.org/10.1007/s00439-019-02062-0.

Milner, Yeshimabeit. n.d. "Action: COVID-19 Open Data by State." Data for Black Lives. https://d4bl.org/action.html. Last accessed December 2021.

Mishra, Rajiv. 2020. "The Digital State: A Tale of Tweets and Foods in Contemporary India." In *Digital Transactions in Asia: Economic, Informational, and*

Social Exchanges, edited by Adrian Athique and Emma Baulch, 156–71. New York: Routledge.
Mitchell, Margaret, Simone Wu, Andrew Zaldivar, Parker Barnes, Lucy Vasserman, Ben Hutchinson, Elena Spitzer, Inioluwa Deborah Raji, and Timnit Gebru. 2019. "Model Cards for Model Reporting." In *Proceedings of the Conference on Fairness, Accountability, and Transparency*, 220–29. New York: Association for Computing Machinery. https://doi.org/10.1145/3287560.3287596.
Murphy, Sharon Ann. 2013. *Investing in Life: Insurance in Antebellum America*. Baltimore, MD: Johns Hopkins University Press.
———. 2021. "Enslaved Financing of Southern Industry: The Nesbitt Manufacturing Company of South Carolina, 1836–1850." *Enterprise and Society*, February 4, 2021, 1–44. https://doi.org/10.1017/eso.2020.78.
Murphy-Abdouch, Kim. 2015. "Patient Access to Personal Health Information: Regulation vs. Reality." *Perspectives in Health Information Management* 12 (Winter). https://www.ncbi.nlm.nih.gov/pmc/articles/PMC4700868/.
Nass, Sharyl J., Laura Levit, and Lawrence O. Gostin, eds. 2009. *Beyond the HIPAA Privacy Rule: Enhancing Privacy, Improving Health Through Research*. Washington DC: Institute of Medicine, Nation Academies Press.
Neyland, Daniel. 2019. *The Everyday Life of an Algorithm*. New York: Palgrave Macmillan.
Nielsen, Rasmus Kleis. 2012. *Ground Wars: Personalized Communication in Political Campaigns*. Princeton, NJ: Princeton University Press.
Nissenbaum, Helen. 2010. *Privacy in Context: Technology, Policy, and the Integrity of Social Life*. Stanford, CA: Stanford University Press.
———. 2011. "A Contextual Approach to Privacy Online." *Dædalus* 140 (4): 32–48.
Nixon, Rob. 2013. *Slow Violence and the Environmentalism of the Poor*. Cambridge, MA: Harvard University Press.
Noble, Safiya Umoja. 2018. *Algorithms of Oppression: How Search Engines Reinforce Racism*. New York: NYU Press.
Nunn, Ryan, Jana Parsons, and Jay Shambaugh. 2020. "A Dozen Facts about the Economics of the US Health-Care System." The Brookings Institution, March 10, 2020. https://www.brookings.edu/research/a-dozen-facts-about-the-economics-of-the-u-s-health-care-system/.
Obermeyer, Ziad, Brian Powers, Christine Vogeli, and Sendhil Mullainathan. 2019. "Dissecting Racial Bias in an Algorithm Used to Manage the Health of Populations." *Science* 366 (6464): 447. https://doi.org/10.1126/science.aax2342.
O'Connor, Bonnie, Fran Pollner, and Adriane Fugh-Berman. 2016. "Salespeople in the Surgical Suite: Relationships between Surgeons and Medical Device Representatives." *PloS One* 11 (8): e0158510. https://doi.org/10.1371/journal.pone.0158510.

Office for Civil Rights (OCR), Department of Health and Human Services. 2013. "Modifications to the HIPAA Privacy, Security, Enforcement, and Breach Notification Rules under the Health Information Technology for Economic and Clinical Health Act and the Genetic Information Nondiscrimination Act; Other Modifications to the HIPAA Rules." 45 CFR 160 and 164.

———. 2020. "Notification of Enforcement Discretion for Telehealth Remote Communications during the COVID-19 Nationwide Public Health Emergency." https://www.hhs.gov/hipaa/for-professionals/special-topics/emergency-preparedness/notification-enforcement-discretion-telehealth/index.html.

Olson, John, and Andrea Phillips. 2013. "Rikers Island: The First Social Impact Bond in the United States." *Community Development Investment Review*, April, 97–101.

O'Neil, So, Colleen Staatz, Emily Hoe, Ravi Goyal, and Eva Ward. 2019. "Data across Sectors for Health Initiative: Formative Evaluation Report." Robert Wood Johnson Foundation. https://dashconnect.org/about-dash/evaluation-and-key-accomplishments-from-dashs-first-five-years/.

Osman, Jena. 2019. *Motion Studies*. New York: Ugly Duckling Presse.

Papanicolas, Irene, Liana R. Woskie, and Ashish K. Jha. 2018. "Health Care Spending in the United States and Other High-Income Countries." *JAMA* 319 (10): 1024–39. https://doi.org/10.1001/jama.2018.1150.

Park, Joohyun, and Kevin A. Look. 2019. "Health Care Expenditure Burden of Cancer Care in the United States." *INQUIRY: The Journal of Health Care Organization, Provision, and Financing* 56 (January). https://doi.org/10.1177/0046958019880696.

Pasquale, Frank. 2015. *The Black Box Society: The Secret Algorithms That Control Money and Information*. Cambridge, MA: Harvard University Press.

———. 2018. "Quantifying Love," *Boston Review*, Fall 2018, 106–13.

Pell, Susan. 2015. "Radicalizing the Politics of the Archive: An Ethnographic Reading of an Activist Archive." *Archivaria*, November, 33–57.

Peterson, Victor, II. 2021. "Improvisation in Poetic Computation." In *Black Archives and Intellectual Histories*, edited by Khwezi Mkhize, Mandisa Haarhoff, and Christopher Ouma. Johannesburg: Wits University Press, forthcoming.

Pierson, Emma, David M. Cutler, Jure Leskovec, Sendhil Mullainathan, and Ziad Obermeyer. 2021. "An Algorithmic Approach to Reducing Unexplained Pain Disparities in Underserved Populations." *Nature Medicine* 27 (1): 136–40. https://doi.org/10.1038/s41591-020-01192-7.

Pink, Sarah, Debora Lanzeni, and Heather Horst. 2018. "Data Anxieties: Finding Trust in Everyday Digital Mess." *Big Data and Society* 5 (1): 2053951718756685. https://doi.org/10.1177/2053951718756685.

Plott, Caroline F., Allen B. Kachalia, and Joshua M. Sharfstein. 2020. "Unexpected Health Insurance Profits and the COVID-19 Crisis." *JAMA* 324 (17): 1713. https://doi.org/10.1001/jama.2020.19925.

Poon, Martha. 2007. "Scorecards as Devices for Consumer Credit: The Case of Fair, Isaac and Company Incorporated." *Sociological Review* 55 (2 suppl): 284–306. https://doi.org/10.1111/j.1467-954X.2007.00740.x.

———. 2009. "From New Deal Institutions to Capital Markets: Commercial Consumer Risk Scores and the Making of Subprime Mortgage Finance." *Accounting, Organizations and Society* 34 (5): 654–74. https://doi.org/10.1016/j.aos.2009.02.003.

———. 2012. "What Lenders See: A History of the Fair Isaac Scorecard." PhD diss., University of California, San Diego.

Poovey, Mary. 1998. *A History of the Modern Fact: Problems of Knowledge in the Sciences of Wealth and Society*. Chicago: University of Chicago Press.

———. 2018. "Risk, Uncertainty and Data: Managing Risk in Twentieth-Century America." In *American Capitalism: New Histories*, edited by Sven Beckert and Christine Desan, 221–35. New York: Columbia University Press.

Porter, Theodore M. 1995. *Trust in Numbers: The Pursuit of Objectivity in Science and Public Life*. Princeton, NJ: Princeton University Press.

Prabhu, Vinay Uday, and Abeba Birhane. 2020. "Large Image Datasets: A Pyrrhic Win for Computer Vision?" *ArXiv* 2006.16923 [Cs.CY]. http://arxiv.org/abs/2006.16923.

Prainsack, Barbara. 2017. *Personalized Medicine: Empowered Patients in the Twenty-first Century?* New York: New York University Press.

Radin, Joanna. 2017. "'Digital Natives': How Medical and Indigenous Histories Matter for Big Data." *Osiris* 32 (1): 43–64.

Raji, Inioluwa, and Joy Buolamwini. 2019. "Actionable Auditing: Investigating the Impact of Publicly Naming Biased Performance Results of Commercial AI Products." *Proceedings of the 2019 AAAI/ACM Conference on AI, Ethics, and Society*, 429–35. https://doi.org/10.1145/3306618.3314244.

Raji, Inioluwa, Timnit Gebru, Margaret Mitchell, Joy Buolamwini, Joonseok Lee, and Emily Denton. 2020. "Saving Face: Investigating the Ethical Concerns of Facial Recognition Auditing." In *AIES '20: Proceedings of the AAAI/ACM Conference on AI, Ethics, and Society*, 145–51. https://doi.org/10.1145/3375627.3375820.

Resnais, Alain, and Chris Marker, dirs. 1953. *Les Statues meurent aussi (Statues also die)*. Paris: Présence Africaine.

Rice, Lisa, and Deidre Swesnik. 2014. "Discriminatory Effects of Credit Scoring on Communities of Color." *Suffolk University Law Review* 46: 935–66.

Roberts, Dorothy E. 2011. *Fatal Invention: How Science, Politics, and Big Business Re-Create Race in the Twenty-First Century*. New York: New Press.

Robinson + Yu. 2014. "Knowing the Score: New Data, Underwriting, and Marketing in the Consumer Credit Marketplace: A Guide for Financial Inclusion

Stakeholders." Washington DC: Robinson + Yu. https://www.upturn.org/static/files/Knowing_the_Score_Oct_2014_v1_1.pdf.
Rothstein, Richard. 2018. *The Color of Law: A Forgotten History of How Our Government Segregated America*. New York: Liveright.
Rowe, Rachel, and Niamh Stephenson. 2016. "Speculating on Health: Public Health Meets Finance in 'Health Impact Bonds.'" *Sociology of Health and Illness* 38 (8): 1203–16. https://doi.org/10.1111/1467-9566.12450.
Ruckenstein, Minna, and Natasha Dow Schüll. 2017. "The Datafication of Health." *Annual Review of Anthropology* 46 (1): 261–78. https://doi.org/10.1146/annurev-anthro-102116-041244.
Rudin, Cynthia, Caroline Wang, and Beau Coker. 2020. "The Age of Secrecy and Unfairness in Recidivism Prediction." *Harvard Data Science Review* 2 (1). https://doi.org/10.1162/99608f92.6ed64b30.
Sanderson, Saskia C., Kyle B. Brothers, Nathaniel D. Mercaldo, Ellen Wright Clayton, Armand H. Matheny Antommaria, Sharon A. Aufox, Murray H. Brilliant, et al. 2017. "Public Attitudes toward Consent and Data Sharing in Biobank Research: A Large Multi-Site Experimental Survey in the US." *The American Journal of Human Genetics* 100 (3): 414–27. https://doi.org/10.1016/j.ajhg.2017.01.021.
Sänger, J., C. Richthammer, S. Hassan, and G. Pernul. 2014. "Trust and Big Data: A Roadmap for Research." In *2014 25th International Workshop on Database and Expert Systems Applications*, 278–82. https://doi.org/10.1109/DEXA.2014.63.
Schiros, Chun G., Thomas S. Denney, and Himanshu Gupta. 2015. "Interaction Analysis of the New Pooled Cohort Equations for 10-Year Atherosclerotic Cardiovascular Disease Risk Estimation: A Simulation Analysis." *BMJ Open* 5 (4). https://doi.org/10.1136/bmjopen-2014-006468.
Scholz, Lauren Henry. 2016. "Privacy as Quasi-Property." *Iowa Law Review* 101: 1113–20.
Schwartz, Paul M. 2009. "Preemption and Privacy." *Yale Law Journal* 118 (5): 902–47.
Shilo, Smadar, Hagai Rossman, and Eran Segal. 2020. "Axes of a Revolution: Challenges and Promises of Big Data in Healthcare." *Nature Medicine* 26 (1): 29–38. https://doi.org/10.1038/s41591-019-0727-5.
Singh, Gopal K., and Stella M. Yu. 2019. "Infant Mortality in the United States, 1915–2017: Large Social Inequalities Have Persisted for Over a Century." *International Journal of Maternal and Child Health and AIDS* 8 (1): 19–31. https://doi.org/10.21106/ijma.271.
Skeem, Jennifer L., and Christopher T. Lowenkamp. 2016. "Risk, Race, and Recidivism: Predictive Bias and Disparate Impact." *Criminology* 54 (4): 680–712. https://doi.org/10.1111/1745-9125.12123.
Slater, M. D., and J. A. Flora. 1991. "Health Lifestyles: Audience Segmentation

Analysis for Public Health Interventions." *Health Education Quarterly* 18 (2): 221–33.

Star, Susan Leigh, and James R. Griesemer. 1989. "Institutional Ecology, 'Translations' and Boundary Objects: Amateurs and Professionals in Berkeley's Museum of Vertebrate Zoology, 1907–39." *Social Studies of Science* 19 (3): 387–420. https://doi.org/10.1177/030631289019003001.

Stark, Peter. 2010. "Congressional Intent for the HITECH Act." *American Journal of Managed Care* 16 (12): SP24–28.

Stepanikova, Irene, Karen Cook, David Thom, Roderick Kramer, and Stefanie Mollborn. 2009. "Trust in Managed Care Settings." In *Whom Can We Trust?*, edited by Karen S. Cook, Margaret Levi, and Russell Harden, 149–81. New York: Russell Sage Foundation.

Strasser, Bruno J. 2019. *Collecting Experiments: Making Big Data Biology.* Chicago: University of Chicago Press.

TallBear, Kim. 2013. *Native American DNA: Tribal Belonging and the False Promise of Genetic Science.* Minneapolis: University of Minnesota Press.

Tanner, Adam. 2017. *Our Bodies, Our Data: How Companies Make Billions Selling Our Medical Records.* New York: Beacon Press.

Taylor, Keeanga-Yamahtta. 2019. *Race for Profit: How Banks and the Real Estate Industry Undermine Black Homeownership.* Chapel Hill: University of North Carolina Press.

Tempini, Niccolò, and Lorenzo Del Savio. 2018. "Digital Orphans: Data Closure and Openness in Patient-Powered Networks." *BioSocieties* 14:205–27. https://doi.org/10.1057/s41292-018-0125-0.

Thomas, Jesse. 2017. "Best Practices in the Implementation of Open Data at a Municipal Government Level." Master's thesis, University of Oregon.

Thomas, Lyn C. 2000. "A Survey of Credit and Behavioural Scoring: Forecasting Financial Risk of Lending to Consumers." *International Journal of Forecasting* 16 (2): 149–72. https://doi.org/10.1016/S0169-2070(00)00034-0.

Thornton, Lauren, Bran Knowles, and Gordon Blair. 2021. "Fifty Shades of Grey: In Praise of a Nuanced Approach Towards Trustworthy Design." In *Proceedings of the 2021 ACM Conference on Fairness, Accountability, and Transparency*, 64–76. New York: Association for Computing Machinery.

Tovino, Stacey. 2016. "Complying with the HIPAA Privacy Rule: Problems and Perspectives." *Scholarly Works*, paper 999. http://scholars.law.unlv.edu/facpub/999.

Traub, Amy. 2014. "Discredited: How Employment Credit Checks Keep Qualified Workers out of a Job." *Demos*, February 3, 2014. https://www.demos.org/research/discredited-how-employment-credit-checks-keep-qualified-workers-out-job.

Trundle, Catherine, and Chris Kaplonski. 2011. "Tracing the Political Lives of

Archival Documents." *History and Anthropology* 22 (4): 407–14. https://doi.org/10.1080/02757206.2011.626777.

Tufte, Edward. 2006. *Beautiful Evidence*. Cheshire, CT: Graphics Press.

Upadhyay, Soumya, Amber L. Stephenson, and Dean G. Smith. 2019. "Readmission Rates and Their Impact on Hospital Financial Performance: A Study of Washington Hospitals." *INQUIRY: The Journal of Health Care Organization, Provision, and Financing* 56. https://doi.org/10.1177/0046958019860386.

Urban Institute. n.d. "Debt in America: An Interactive Map." Last updated March 31, 2021. https://apps.urban.org/features/debt-interactive-map.

Vogenberg, F. Randy. 2009. "Predictive and Prognostic Models: Implications for Healthcare Decision-Making in a Modern Recession." *American Health and Drug Benefits* 2 (6): 218–22.

Vyas, Darshali A., Leo G. Eisenstein, and David S. Jones. 2020. "Hidden in Plain Sight: Reconsidering the Use of Race Correction in Clinical Algorithms." *New England Journal of Medicine* 383 (9): 874–82. https://doi.org/10.1056/NEJMms2004740.

Waldby, Catherine. 1998. "Medical Imaging: The Biopolitics of Visibility." *Health:* 2 (3): 372–84. https://doi.org/10.1177/136345939800200306.

Ward, Paul Russell. 2017. "Improving Access to, Use of, and Outcomes from Public Health Programs: The Importance of Building and Maintaining Trust with Patients/Clients." *Frontiers in Public Health* 5 (22). https://doi.org/10.3389/fpubh.2017.00022.

Washington, Harriet. 2006. *Medical Apartheid: The Dark History of Medical Experimentation on Black Americans from Colonial Times to the Present.* New York: Doubleday.

———. 2021. "Medical Apartheid Goes Viral." Lecture presented at Racism, Medicine, and Bioethics: Learning from the Past to Ensure a Healthier Future, Tuskegee University National Center for Bioethics in Research and Healthcare, and Harvard Medical School Center for Bioethics, February 3, 2021. https://bioethics.hms.harvard.edu/events/black-history-month-event-series.

Wei, Yanhao, Pinar Yildirim, Christophe Van den Bulte, and Chrysanthos Dellarocas. 2016. "Credit Scoring with Social Network Data." *Marketing Science* 35 (2): 218–33.

Wells, Jonathan Daniel. 2020. *The Kidnapping Club: Wall Street, Slavery, and Resistance on the Eve of the Civil War*. New York: Bold Type Books.

Wilkinson, Mark D., Michel Dumontier, IJsbrand Jan Aalbersberg, Gabrielle Appleton, Myles Axton, Arie Baak, Niklas Blomberg, et al. 2016. "The FAIR Guiding Principles for Scientific Data Management and Stewardship." *Scientific Data* 3 (1): 1–9. https://doi.org/10.1038/sdata.2016.18.

Williams, Simon J., Catherine Coveney, and Robert Meadows. 2015. "'M-Apping'

Sleep? Trends and Transformations in the Digital Age." *Sociology of Health and Illness* 37 (7): 1039–54. https://doi.org/10.1111/1467-9566.12283.

Williams, Winfred W., Joseph W. Hogan, and Julie R. Ingelfinger. 2021. "Time to Eliminate Health Care Disparities in the Estimation of Kidney Function." *New England Journal of Medicine* 385 (19): 1804–6. https://doi.org/10.1056/NEJMe2114918.

Wittgenstein, Ludwig. 2010. *Tractatus Logico-Philosophicus*. New York: Harcourt, Brace and Co.

Woolgar, Steve, and Daniel Neyland. 2014. *Mundane Governance: Ontology and Accountability*. Oxford: Oxford University Press.

Wu, Tim. 2016. *The Attention Merchants: The Epic Scramble to Get Inside Our Heads*. New York: Knopf.

Zafar, S. Yousuf. 2016. "Financial Toxicity of Cancer Care: It's Time to Intervene." *JNCI: Journal of the National Cancer Institute* 108 (5). https://doi.org/10.1093/jnci/djv370.

Zafar, S. Yousuf, and Amy P. Abernethy. 2013. "Financial Toxicity, Part I: A New Name for a Growing Problem." *Oncology* 27 (2): 80–149.

Zeldin, Cindy, and Mark Rukavin. 2007. "Borrowing to Stay Healthy: How Credit Card Debt Is Related to Medical Expenses." Demos. https://www.aecf.org/resources/borrowing-to-stay-healthy/.

Zerilli, John, Alistair Knott, James Maclaurin, and Colin Gavaghan. 2019. "Algorithmic Decision-Making and the Control Problem." *Minds and Machines* 29 (4): 555–78. https://doi.org/10.1007/s11023-019-09513-7.

Zuboff, Shoshana. 2015. "Big Other: Surveillance Capitalism and the Prospects of an Information Civilization." *Journal of Information Technology* 30 (1): 75–89. https://doi.org/10.1057/jit.2015.5.

———. 2019. *The Age of Surveillance Capitalism: The Fight for a Human Future at the New Frontier of Power*. New York: PublicAffairs, Perseus Book Group.

Index